Corporate Television

Corporate Television

A Producer's Handbook

Ray DiZazzo

Photography by F. Martin

Focal Press
Boston London

Focal Press is an imprint of Butterworth Publishers.

Library of Congress Cataloging-in-Publication Data

DiZazzo, Raymond.
Corporate television : a producer's handbook / Ray DiZazzo; photography by F. Martin.
p. cm.
Bibliography: p.
Includes index.
ISBN 0-240-80023-0
1. Video recordings—Production and direction. 2. Television—Production and direction. 3. Industrial television. 4. Television in management. I. Title.
PN1992.94.D58 1989
784.55'57—dc19 88-35485
CIP

British Library Cataloguing in Publication Data

DiZazzo, Ray
Corporate television : a producer's handbook.
1. Television programmes. Production.—Manuals
I. Title
791.45'0232

ISBN 0-240-80023-0

Butterworth Publishers
80 Montvale Avenue
Stoneham, MA 02180

10 9 8 7 6 5 4 3 2 1

Printed in the United States of America

Contents

Preface

Over the nearly ten years that I have been involved in producing corporate television programs, I have become very familiar with the numerous activities and elements that make up the television production process.

When I first entered the field, however, the opposite was true. Although I had virtually no background in the business, I quickly found myself in a position of control—a situation not too unusual in corporate television.

Although this turned out to be a very challenging and rewarding period in my career, it was also a very uncomfortable one. I am sure now that the most difficult factor for me during that time was simply not having a framework or an overview of the entire process—a comfortable feel for how all the pieces fit together and, more important, how I fit into that process as a producer.

With that in mind, I have written this book primarily for the student and the new producer, although others in the field, including new corporate television clients, will also find valuable information here. In addition, I have purposely chosen *not* to make this an in-depth reference book. A *framework* is what I have attempted to construct—a brief, concise look at all the elements that go into making a corporate television program and how they are interrelated from a producer's point of view.

Those elements fall into five general categories: writing and design, preproduction, production, postproduction, and evaluation. In addition to these areas, I have also included information on two other important subjects—client relations and future survival in corporate television.

The student or new producer should consider this book a first step. The in-depth material not covered in these pages is readily available in colleges, bookstores, and libraries, and I would recommend further exploration of each of the areas just mentioned once this book has provided the foundation on which that knowledge can be built.

In addition to gaining technical knowledge, I would also urge corporate television newcomers to resign themselves at the outset to mustering plenty of patience and persistence. Like any other profession, corporate television cannot be learned overnight. Even for the brightest students, it requires a good deal of time and energy. In addition, breaking the ice—finding that first job in the business—can be difficult.

For those who persist, however, the rewards can be great. The corporate television field, like video, the electronic process on which it is built, is part of a cultural explosion taking place in modern business. That explosion is based on the knowledge that communications on a vast and immediate scale are crucial to a company's health and sustained growth. Add to this the realization that the quickest and most effective way to communicate a consistent message to large numbers of employees at once is through the use of corporate television, and it becomes apparent just how bright and exciting the future is.

Ray DiZazzo

Acknowledgments

Although many people have been of great help to me throughout my corporate television career, I would like to dedicate this book to those few people who were instrumental in actually getting that career off the ground.

I am extremely grateful to Joanne Terry and Art Stoebe for their acute perception and willingness to let me explore new territory. I am grateful, also, to Bill Alexander and Dick Jones because they took the greatest risk on the green newcomer, and to Glenn Hunter, who helped me through those first extremely tough years with his generous concern, constant support, and knowledgeable guidance.

Speaking of guidance, let me not forget Allan Amenta, an excellent writer, friend, and teacher, and Ned Rodgers, who taught me what directing television was all about.

Last but not least, I am forever grateful to my wife, Patti, and my two children, Sunday and Sean—the unseen ones—who provided that all-important *family* support on the many occasions it was sorely needed.

1 Corporate Television and the Corporate Producer

TELEVISION: PUBLIC OR PRIVATE MEDIUM?

Millions of Americans watch television every day. In most cases, their reasons for breaking out the potato chips, flopping into a comfortable chair, and turning on "the tube" can be summed up in one simple word: *entertainment*. Game shows, sports programs, the daily soap operas and evening sitcoms—all are typical examples of programming specifically produced to satisfy this need for entertainment.

But there are also other types of programs on television—programs that have more to offer than straight entertainment. Most network and many cable channels, for instance, also carry documentaries. Then there are the daily news shows and provocative specials. The educational and cultural programming aired on Public Broadcasting System (PBS) stations is still another example. In each of these cases, although the programs often are very entertaining, they are also teaching or at least *informing* the *public* about something.

CORPORATE TELEVISION: THE PRIVATE MEDIUM

Corporate television—programming produced specifically for *private* use in business and industry—is very similar to this type of public television. Corporate television is written, produced, directed, and edited using the same methods, much of the same equipment and technology, and often the same people. Like public TV, it can also be very entertaining, although its primary goal is to inform.

Besides these technical and aesthetic similarities, TV broadcast to the public and corporate television have still another common thread. It is one that is sometimes overlooked (at least in the corporate world), although it is the basic reason *both* types of programming exist.

Profit: The Bottom Line

In the final analysis, both public and corporate television programming is produced for *profit*. In television broadcast to the public, this profit is achieved by entertaining a large number of people, thereby gaining their loyalty as viewers. This loyal public viewership will watch certain programs regularly, be exposed to advertising for the sponsor's products, and consequently buy those products. A profit is made; the writers, producers, directors, and others involved are paid for their services; more programming is financed for production; and so the cycle continues.

In corporate television, profit is achieved in a less tangible but just as important way—by influencing the attitudes and behaviors of a *private* viewership (company employees) in a way that will positively (i.e., profitably) affect the company's performance.

Communicating Information

Achieving this type of corporate profit through television almost always involves communicating three types of information: training, motivational, and news.

Training information can teach employees specific ways to sell a product or deal professionally with customers on the phone.

Motivational information can emphasize how high a company's productivity was for the previous quarter and encourages employees to achieve even better results in the next quarter.

News information is normally designed to keep employees up to date on developments within the company. This helps maintain good morale and high productivity.

CORPORATE PROGRAMMING: FORMATS

Just how does a corporate television department get these training, motivational, and news messages out to employees? For the most part, it's done in two ways: video cassette distribution and private broadcast.

Video Cassette Distribution

In the case of cassette distribution, programs are written, shot, edited, and finally duplicated to be sent to employee locations where a television and a video cassette recorder (VCR) are kept. A local presenter (usually a supervisor) receives the cassette through company mail and, when time permits, gathers employees in a conference room. She shows the program and perhaps answers questions or hands out written material as support.

Another example of the use of cassette distribution is during formal company training classes. In this case, the program is often designed in advance as a part of the course's curriculum development. The cassette is kept on site by the training instructor and used in a TV/VCR at the appropriate time during each class.

Private Corporate Broadcast

Private broadcast, on the other hand, has the advantage of being able to air "live." In this case, programming does not have to be recorded on videotape in advance. It can be transmitted as it is happening via microwave or satellite links. The signals are picked up by company television sets connected to receiver feeds at each employee location.

Typical corporate broadcasts might include important messages from the executive ranks, company news, or training or motivational material on new products. Live corporate broadcasts often allow audience members to call in and interact with those in the studio.

Prerecorded programming on video cassette can also be broadcast in this way instead of being sent to the field and played on VCRs.

Other Methods

In some companies, other forms of employee communications, such as slide programs, may also fall under the responsibility of the corporate television department. Often, however, these are left to different departments that specialize in such activities.

PROGRAM FINANCING

In the past, many corporate television departments worked with internal budgets for productions. They were allocated a yearly dollar amount and used that money as they saw fit to produce programs for various departments within their companies. This arrangement was convenient for two reasons. First, it provided departmental security: Money was guaranteed each year. Second, decisions about how to use that money were kept internal.

These days, however, what are called *chargeback systems* are more common. In a chargeback system, the television department has a very minimal internal production budget, or none at all. Its funding comes solely from the budgets of the departments that come to it for programming. In other words, it *charges back* the costs associated with each production to the department requesting it.

The chargeback system of operating is an economical one for companies in general, but it often poses special challenges for corporate producers. Just one example involves client input. When the television department is paying the bills, clients are apt to be very willing to go along with the producer's decisions. When *they* are paying, however, clients can be much more demanding about scripts, actors, locations, allocation of budget monies, and so on.

The solution to problems of this type usually lies in the client–producer relationship. The producer should be doing work of high enough quality for low enough prices to keep key departments in the company convinced that they are in competent professional hands. By the same token, the client should trust the producer enough to allow him to exercise the proper amount of control over the project.

TELEVISION'S PLACE IN THE CORPORATE WORLD

On the organizational chart, the corporate television department often appears as part of a general office staff group, such as public affairs, human resources, or internal/external communications. This type of reporting structure allows it to maintain a company-wide scope in its programming because these departments typically work on a company-wide basis. It also allows for the possibility of producing some programs for the public as in-kind or community services. These usually include public service announcements (PSAs) to support charities or other worthy community causes.

Aside from these factors, a public affairs type of reporting structure is also advantageous because it keeps the television group in close proximity to executive circles. It is in these circles that much of the company's news and other important information will either originate or surface for dissemination.

The danger of a general office reporting structure is the possibility that the television department may become too involved with executive "wheel greasing" and begin to ignore the people who really need its support—employees on the front lines.

If a periodic check of the production schedule begins to show a substantial number of such executive projects, some departmental self-examination is probably in order.

However it is structured, the corporate world—especially in very large companies—can be a challenging place for any television department. This is because it is primarily a world of high finances, business protocol, three-piece suits, and lots of politics. To the people in charge, it is a world of television only secondarily. The wise producer learns this quickly and keeps these priorities in clear perspective when dealing with clients, bosses, and company executives.

In general, the corporate world is also more conservative than the world of feature films or network television. This can sometimes create a frustrating situation, because the need for high levels of creativity is also an inherent part of the corporate television process. Script writing is creative work. So is producing and directing. Sometimes, however, the people who possess these creative talents (those the producer most hire) are the ones who find it most difficult to operate in a highly structured, conservative corporate environment. This conservative–creative conflict is just one of the many challenges the corporate producer must deal with on a continual basis.

THE CORPORATE PRODUCER

The word *corporate* is used in this book to refer to business or industry in general. *Business television* and *industrials* are two common synonyms for *corporate television.* The corporate producer, then, is any person who produces programs for use in this general business world. In this book, he or she is assumed to be on staff in a large company, as is a very common situation.

On-Staff Producers

The on-staff corporate producer's job is typically a middle-level management position. In some cases, it involves supervising support employees such as staff associate producers, secretaries, or production assistants. In almost all cases, it involves hiring and supervising freelancers such as writers, directors, camera operators, and editors.

Freelancers

But not all producers are on staff. A corporate producer can be a freelancer himself, working for an assortment of companies on an independent, per-project basis. Many freelancers work in both the entertainment and the corporate world at different times, depending on where they can find work.

Producer "Musts"

Whether the producer is a staffer or a freelancer makes little difference when it comes to producing programs. In either case, the corporate producer must be a jack-of-all-media-trades. His primary skills must lie in the areas of project and people management. He must be able to closely guide the many elements of the production process and supervise the creative people who conceive and carry out those elements.

This is not to say that the producer is not creative herself. Not only must she be extremely creative in the expected areas—writing, directing, and editing—but she must also be creative in her use of the company's time, money, and human resources.

In order to guide and supervise television productions effectively, the producer must have a working knowledge of all the functions involved in that process. He must know what it is like to sit in front of a word processor staring at the blank screen, under pressure to have a shooting script by the next morning. He must know all the labors of actually getting that script written, revised, approved, and placed in the hands of a capable director. He must be well aware of and able to use the human skills involved in questioning a difficult executive or client, and he must be skillful at weeding out the many unusable answers he will get from those that are of value.

The producer must also know the director's job intimately. She must be closely in touch with the aesthetic process of motion picture photography, the needs and temperaments of actors, and the wants and needs of the crew of people who will help her bring the pages of the script to life on the television screen. She must know how to convince a client, tactfully, that his opinion is wrong, how to tell the company president to straighten up for the camera, and how to make valid creative decisions under extreme pressures. She must be able to break down a script, write a shooting schedule, set a key light, create a budget, and develop a shot list.

But the producer's "musts" don't stop there. He must also know the postproduction process. He must understand the frustration an on-line editor is feeling when he says the time code on the window dubs doesn't match the original footage. He must know the feeling of sitting in a dimly lit room staring at television monitors and trying to make a coherent scene out of footage that was shot with no thought given to screen direction, proper focal lengths, or continuity. He must know the frustration of trying to read script notes that don't make sense or trying to use an edit system that keeps slipping frames.

As if this weren't enough, the television producer must also have many other skills. If she is a staffer, she must be a politician able to defend her department to executives who know little and care less about the television production process. She must be an accountant able to manage large sums of money effectively. She must be an actor and a psychologist, and must know how to get what she knows is right, even from very persuasive people who will continually tell her she is dead wrong.

Not only must the producer understand all these things, but in many cases he must also be able to *do* them. He must be an accomplished script writer and a seasoned director. He must also be able to grip and gaff and light and sit in an edit bay and physically cut his own footage. This is because the corporate producer is often a one-person crew, without the luxury of bringing in people to staff his various crew positions.

Above all, the successful producer must have a sense of what *works.* Although this may sound like the easy part, it is probably the most difficult. The ability to know what works on the screen and the confidence to fight for it do not come overnight. They are the product of time, experience, success, failure, and plenty of plain old hard work.

The Reward

What is the payoff for all these "musts"? On one level, of course, there is the paycheck, which can be quite comfortable if the producer is good at what she does—$40,000 to $50,000 a year is fairly common, and some producers make much more. On another, perhaps more important, level, there is the self-satisfaction that comes when the producer sits in a screening room and watches the final program.

In this case, the payoff is a series of shots that cut with seamless precision; actors who bring their roles to life with unquestionable sincerity; a script that flows with simple eloquence from one scene to the next; a sparkle of excitement in the cient's eye; the knowledge that hundreds or even thousands of employees will be helped; and, most of all, the self-satisfaction that comes from having supervised, nurtured, guided, and helped make a reality every element playing on the television screen.

Does this sound like the job for you? If so, read on and we will follow the production process from start to finish just as it is carried out in many companies today.

As you read this book, keep in mind that I have purposely written it assuming an ideal situation—that the producer has the luxury (i.e., the budget) to hire the writer, director, assistant director, and other people he needs. This may or may not be the case in your company or the companies you work for. If it is not, as you read, simply place yourself in the role of the person *doing* the work as well as the one supervising it.

For the sake of clarity and simplicity, I have included the usual elements of the production process in the order in which they normally happen. As with hiring personnel, however, the usual may not always be what happens in the real world. In some situations, for instance, a program will be shot before the script is written, or no script notes can be taken, or the director will also be the camera operator. There will also be times when off-line editing is skipped, when no audio sweetening is necessary, when preproduction planning amounts to a frantic few hours of phone calls, when the client is also the program's host, and so on.

In corporate television, anything is possible. It is often the needs and resources at hand, rather than the theoretical norm, that dictate exactly how a program will be produced.

And that brings up another very important trait the corporate television producer must possess—*lots* of flexibility.

part
one

Writing and Design

2 The Program Needs Analysis

OVERVIEW

A client comes to the corporate television department with a request for a program. The first step in producing that program might seem to be writing, but it is not. Nor is it research. Rather, the first step is an analysis of the client's needs and the intended audience. From these facts come a decision as to whether a television production is the answer to those needs and, if it is, what the design of the program should be.

If the decision is to proceed with a program, the next step is to involve a writer. Then comes research, development of a content outline, a treatment, a first draft of the script, revisions as needed, and finally a client-approved shooting script.

Client approvals are built into this process all along the way to ensure that the project moves in the right direction—according to the design laid out in the beginning—and that there are no surprises the day before (or, worse, the day *after*) production.

THE CLIENT CONTACT

> *This is Ellen Prescott in Safety. We've got a real problem that we need some help on—back injuries. How bad a problem? During the past year we figure the company has lost over $90,000 in direct medical expenses due to back problems—and that doesn't count the productivity loss. We need to get a video out to these people to educate them on how to lift the right way and maybe some steps on how to take care of their backs.*

The request is a typical one. In this case, it takes the form of a phone call from a prospective client. But it could just as easily be a letter, a memo, or a note from your boss to call someone in the training or public affairs department. The request can come in any number of ways; the point is that it usually comes from someone who needs help communicating a message and sees videotape as the perfect solution—often, the *only* solution.

In many cases, a videotape does indeed turn out to be the best tool for the job. For the moment, however, let's suppose it's not, but you make the program anyway.

The Scenario of the Unneeded Videotape

The scenario of the unneeded videotape usually goes something like this: The tape is produced at a cost of hundreds or even thousands of times the cost of solving the problem in some simpler form. The videotape doesn't do what it was really intended to do because it wasn't the right tool for the job in the first place. It may be used a little at first, but then it is shelved forever. The client's objective is met in one sense—she took some action to try to solve the problem—but the people the tape was made for are left out in the cold, and the problem persists.

And what about the corporate television department that produced it? It won't be long before some executive begins to ask a very good question: What are all these videotapes doing sitting on shelves, costing lots of money, when none of the problems around the company are being solved?

How do you avoid this type of mistake and, at the same time, obtain the information you need to design a successful program? You take the first logical step in producing a corporate television program, a simple front-end analysis—a brief, highly focused look at the basics of what the client wants to do, who her audience is, how soon she needs it done, and whether it will really benefit her department and the company as a whole.

In other words, you, the producer, set up an initial client meeting. From the facts you gather in that meeting, you develop a *program needs analysis* (PNA).

The Initial Client Meeting and the Project Overview

Normally, the best way to begin a meeting with a new client is with a brief explanation of the entire production process and the roles each of you will play in it.

Although this may sound like a long and overly detailed way to open a first meeting, there are several good reasons for going through it. The first is simply to make the client aware of the intricacies of the process and to demonstrate that you, the producer, are knowledgeable enough to guide it to a successful finish. The second, and perhaps more important, is to make the point that the client's *support and input,* in the form of prompt approvals and technical advice, are crucial at every step along the way.

This explanation should follow introductions and perhaps some brief discussion of the problem, since that's what the client will probably want to talk about first. The following is a loosely scripted monologue showing in a general way how that initial explanation should be conveyed.

> *At today's meeting, I'm primarily interested in gathering some basic facts from which I'll develop a two- to three-page program needs analysis. This will lay out the facts surrounding the need for your program and what it's intended to accomplish, and examine the audience it's intended for. When it's completed, I'll send it to you for approval.*
>
> *Provided we decide a videotape is the right tool for the job, and the needs analysis is approved, I'll then hire a writer, who will want to meet with you again to gather the content needed to develop the script [unless, of course, you, the producer, will be doing the writing yourself]. During this next meeting, or following it, the writer may want to visit the job site, meet with employees or other content experts, and do whatever other research is necessary to gather all the facts.*
>
> *Following this period of research, the writer will develop a* content outline, *which organizes and structures all the facts, and a* treatment, *which is a brief visualization of what the program will look like once it's produced. Again, these will be sent along for your approval.*
>
> *The writer will then develop a first draft of the script. Yet again, it will be sent to you for approval or revision as needed. In the end we'll have an approved* shooting script.
>
> *At this point, the program will be ready for preproduction. I'll hire a director and probably an assistant, whose jobs will be to line up all the necessary details for actually shooting the program. These include scouting and finalizing locations, hiring a crew, auditioning and hiring actors, locating props, developing a shooting schedule, and lots of other work you as our client can be very helpful with.*
>
> *Following this will come production—the shoot itself—during which it's very important that you and a content expert be present if at all possible.*
>
> *After shooting the material, I'll schedule the program into postproduction. In this phase, a* rough cut *will be developed. This is a work print of the program, which probably won't include some elements such as titles, music, sound effects, and so on. It will, however, be an accurate assembly of all the footage into a version you can approve as is or suggest changes to.*
>
> *Once the rough cut is approved, the program will go into the*

final editing process, during which it will be polished and have music and any necessary sound effects added.

At this point, your program will be complete and ready for distribution.

PNA INFORMATION: KEY ELEMENTS

Following this overview, you'll want to get the information you need to write the PNA. PNAs may vary in form and content depending on your company or departmental needs. The key elements of a PNA, however, are as follows:

1. A brief, concise statement describing the purpose for the videotape. This should be stated in terms of a *problem*—the core issue the program will attempt to resolve.
2. Any background surrounding that problem—for example, a brief history of any other material previously developed or used to try to solve it, why the problem exists, how it is damaging the company, and so on.
3. A thorough analysis of the audience, including any demographics available, their feelings about the problem, their dislikes or prejudices, and their level of knowledge on the subject.
4. A group of specific, quantifiable objectives for the program, stated in such a way that the audience could actually be tested on them after viewing the program.
5. How the program will be used—for example, on a stand-alone basis, with a presenter to answer questions, with written support material, and so on.
6. A brief summary of the benefits and/or drawbacks of producing a program on this subject.

The following example should illustrate these points.

PROFESSIONAL CUSTOMER SERVICE:

How It's Done

A Complete Program Needs Analysis

March 6, 1989 Project 513

Subject:	Customer Service
Working title:	''Professional Customer Service . . . How It's Done''
Client:	Donna Mellertin, Customer Service Administrator, (818) 555-2354
Content experts:	Donna Mellertin (as above) Marty Walsh, Customer Service Analyst, (818) 555-6649
Deadline:	December 14, 1989

PURPOSE

Customer service representatives are not skilled in the modern techiques of handling customers. As a result, customer complaints are high, productivity is lagging, and our customer base appears to be shrinking. This program would teach these representatives the proper customer-handling skills to alleviate these negative trends.

BACKGROUND

To date, two steps have been taken to attempt to solve this problem. On-the-job instructional material has been handed out to all customer service representatives outlining the proper steps for handling customers, and supervisory meetings have been held to provide additional support.

Although these steps have undoubtedly been of some help, they have not been totally effective, for several reasons. It appears that few employees, for instance, actually take the time to read and absorb the on-the-job instructional material. In addition, supervisors are not skilled in training employees in customer-handling skills. Although they are often proficient at passing on what has worked for them in the past, this isn't always what needs to be taught at the present.

The result of all these factors is a continuing problem, which has been damaging the company for the past several years.

AUDIENCE

Demographics

The audience is approximately 85% female and 15% male. Nearly all audience members are high school graduates, and approximately 10% have college degrees. The average age is 23 with a few members as young as 18 and slightly more as old as 41. The ethnic mix is approximately 70% white, 22% black, and 6% Hispanic, with the remaining 2% made up primarily of Asian Pacifics.

Attitudes

Employees who deal with customers for extended periods seem to take a somewhat hardened attitude when it comes to proper customer handling. This is especially true for angry or unreasonable customers. This insensitive attitude appears to be the case with our audience.

By this same token, employee attitudes about the work environment, supervision, company policies, and so on seem to be fairly positive.

What employees do not appear to realize is that proper customer handling has a direct relationship to job satisfaction. A program developed along these lines should explore this relationship as a possible theme.

Knowledge

Employee knowledge levels in the area of customer handling will be fairly high when it comes to self-taught actions, but low in the area of proven techniques. It should be noted, however, that these two seemingly different ways of dealing with customers are often not that far apart. It appears that if audience members are shown how their own personal techniques are similar to those being taught, and why certain differences exist, they will be receptive to giving the new techniques a try.

Overall

In general, this audience should be fairly receptive to a program on customer handling, since it can mean more job satisfaction. Care should be taken, however, not to talk down to them, since they are on the firing line daily and, in a sense, are writing the book on the subject as they work.

OBJECTIVES

Having viewed the proposed program, the audience would be able to:

1. List the following three proven customer-handling techniques:
   a. Express genuine concern for the customer's needs or problems.
   b. Make every effort to accomplish what the customer wants, or follow up as needed.
   c. Let the angry customer vent his or her frustration without interrupting.
2. State the main reason that proper customer handling techniques are personally beneficial to them:
   a. Handling customers properly means less frustration and therefore more job satisfaction.

Note: These objectives should be considered approximately 50 percent instructional and 50 percent motivational.

UTILIZATION

This program would be utilized in conference room environments with approximately five to fifteen viewers present at a showing. Although the program should be of a stand-alone nature, a supervisor would normally be present to actually show the program and answer any questions employees may have.

EVALUATION

This program will be evaluated with the use of a questionnaire designed by the television department and the client. The questionnaire will be distributed at initial viewings attended by the producer and the client. Results will be analyzed and a brief report presented to both the television department administration and the client.

SUMMARY

A production of this nature appears to be a sound investment, for several reasons. First, customer-handling techniques are best learned when seen, heard, and acted out. Therefore, short vignettes demonstrating the proper techniques--and, perhaps, the improper ones as well--would be an ideal way to teach these skills.

Second, although exact figures would be hard to nail down, the use of proper customer-handling techniques will mean a definite increase in profits and a reversal of the loss of the company's customer base.

Finally, this subject is virtually timeless. As a result, the shelf life of this program will be five to ten years.

PNA HINTS

A few hints that may help during your PNA meeting relate to your objectives and your audience. Both are extremely important to the design of the program. Often, however, though unintentionally, a client will not be much help with them.

An Objective Translation

Your job as a producer is to help translate what the client is saying into the information you need. As an example, returning to our possible program on back problems, your client may say something like, "I want the tape to make my people understand how improper lifting can hurt their backs." This is a good general description, but it is much too broad to serve as an instructional objective. How, for instance, would anyone really know if the audience "understands"? A more concise translation of that statement might look something like this:

1. Having viewed the proposed program, the audience will be able to:
   a. List three elements of improper lifting:
      - Bending at the waist
      - Twisting
      - Keeping the center of gravity away from the body

This is a much more definable objective, one you could actually use to test the audience on the effectiveness of the program. The key to making this and other instructional objectives definable is the use of specific action words like *list, state,* and *recite* instead of vague or passive words like *know* or *understand.*

Motivational objectives, on the other hand, are typically less instructional and, for this reason, usually can be stated less specifically. For example, if your program will simply attempt to motivate employees to better attendance, they probably do not need any specific instruction. In this case, objectives that state that they will "feel better" about coming to work or "be motivated" to maintain better attendance habits are acceptable.

Audience Interpretation

Frequently a client will misinterpret audience feelings on a given subject either out of ignorance or simply because she wants her people to look good in your eyes. She may, for example, say that audience opinion of a safety program on safe lifting would be positive because they are all "great people" and it is information that will be helpful to them. This may be true, but it may also be true that they have had it up to their eyeballs with memos and policies on safety. Thus, their first reaction to the program may be, "Oh, no. Here we go again!"

It's important to get at the truth on these issues because a program aimed at an audience with a positive attitude on a subject would probably have a totally different design than one aimed at an audience with a negative attitude.

Again, getting at the heart of these issues is one of the producer's key tasks because she will be guiding the program's development throughout the entire process.

THE DECISION

Once you have developed the PNA, it's decision time. Decision making is another key producer function. Should a program be made on the subject or not? The answer to that question depends on a number of factors. Here are a few the PNA should help flush out:

1. Is the problem best solved with the help of visual aid, or could it be just as well done on paper?

   Example: Instruction on the thirty-five dial positions of a circuit analyzer would be just as well done on paper because there is very little movement involved that would require motion pictures, and the information is probably too detailed to retain after one viewing.

2. Is what the client wants achievable on videotape, or is it too detailed or complicated?

   Example: A company expense policy with many facets or alternative actions would probably be too detailed and boring to present in a video program.

3. Is the time frame achievable? (The typical corporate television program takes about two to three months to produce from start to finish.)

   Example: A fully designed and scripted program on customer relations is not a two-week project. An interview with the company president, however, could be done virtually overnight.

4. Is the subject matter changing or dated? Or will it remain as is long enough to justify the expense of the program?

   Example: If a program is to be produced on a new work uniform, the uniform should be *finalized* in terms of colors, patches, ordering, and all other aspects *before* the message is committed to videotape. In addition, if the uniforms are experimental and may be phased out in a few months, the choice of a videotape may be a mistake.

5. Generally speaking, does the importance of the subject and/or the audience size justify the expense of the production?

> *Example:* In a company of thousands, a program for, say, thirty people on the subject of proper coffee breaks would be questionable. On the other hand, if it is three *thousand* people and the loss of productivity is costing the company $200,000 per year, a program on proper coffee breaks would definitely be a good investment.

As you can see, then, positive answers to questions like these can confirm the need for your program. In that case, if you are not writing it yourself, your next step is to hire a writer.

3 The Treatment and Content Outline

THE WRITER

When you're hiring a writer, do not be too heavily influenced by résumés. Almost all of them look impressive, but good corporate writers are few and far between.

Also be on the lookout for the unprofessionals—the writers who, probably because of feelings of insecurity about their writing abilities, believe they have to look and act the part of the eccentric. They typically dress badly, show up late for meetings, miss deadlines, and tend to be overly protective of what they have written. Many no doubt also feel they are doing their "real" work after hours and are only writing corporate scripts as a means of earning a living until they are recognized by the studios. These writers are generally unable to function well in a corporate environment and on top of that, are often not very good at writing either.

Instead, keep your eye out for the real corporate professionals—the writers who dress for a business environment and are well equipped to talk to and work with executives and upper-level management people. They write well, they are open to constructive criticism, and most of all they are working in corporate television because they *want* to.

If you're just getting started, you can often find these people by word of mouth and interview. Call other corporations or producers in the area and ask for a name or two. Most producers will be more than willing to share, and you will be starting out with a person you know has recently satisfied someone else in your position.

Payment for writers varies widely and can be handled in any number of ways. As a ballpark figure, you can assume a good, experienced writer will cost around $1,500 to $2,000 per script, often payable in three phases: one-third as an advance, another third after delivery of the first draft, and the final third on delivery of the approved script. Built into that payment schedule, normally, are two or three revisions after submission of the first draft.

Again, however, this is only a starting point. It can vary depending on a number of things. How immediate is the deadline, for instance?

Will the writer have to drop everything else she's doing and burn the midnight oil on your project? How complicated is the project? Will it require extensive research or maybe just a quick meeting or two? How good is the writer? Is she the type who delivers a first draft so good that all it needs are minor revisions? Or does she require a lot of your time to help shape the script into what you want? Another important consideration, of course, is the budget—either yours or the client's. You may have only a small amount of money to offer no matter what the circumstances. In this case, the amount of payment simply becomes negotiable.

THE DESIGN AND CONTENT/TREATMENT MEETINGS

Program Design and the PNA

Once you have hired your writer, the next step is to share the PNA with her and discuss the program's design.

Just how does the PNA help dictate design? Let's consider a few brief examples. As you will recall, some of the key factors brought out in the PNA focused on the following areas:

1. Audience demographics

   Example: If the demographics show that the audience is made up primarily of highly educated females, a sophisticated female on-camera hostess or voice-over narrator might be a better choice than a male.

2. Audience knowledge

   Example: If the subject of the program is how to trouble-shoot on an electrical circuit, and the PNA shows the audience knows absolutely nothing about electrical current, you would probably want to start the program with a series of basic instructional definitions. Or this might warrant consideration of more than one program—first, a basic electrical overview, followed by an advanced program on trouble-shooting. Written supporting material would also be useful in this situation.

3. Audience attitudes

   Example: If the subject of the program is one the PNA shows the audience may tend to be cynical or suspicious of (major job or company changes, for instance), a documentary format with heavy use of interviews and testimonials will probably have much more credibility than, say, a role-play—especially if those testimonials come from people the audience members respect.

4. Objectives

   Example: If the objectives require that audience members be able to demonstrate three new work functions, the program should probably center around visual demonstrations of those work functions.

As with many other areas covered in this book, these examples touch on just a few of the many considerations that may exist in your company or department. They should, however, provide a basis for the relationship between the PNA and program design.

Content and/or Treatment?

With the design information firmly established in the writer's mind, your next step would be to arrange a meeting between her, the client, and possibly yourself. The purpose of this meeting is both to introduce the writer to the subject and to begin gathering the information needed to develop a content outline, a treatment, or both.

As we discussed earlier, this may expand into a number of meetings and extensive research, including visits to job sites, employee interviews, and so on. Or it may take only one meeting, from which the writer leaves knowing everything she needs to dig in. Whatever the situation, the writer should be able to return to you in roughly one week with either a content outline or a treatment.

THE CONTENT OUTLINE

Most writers develop some sort of outline as the research they are doing on a project begins to take shape. This may be a simple list of bulleted statements on tearsheets, or it may be a formal, extensive, typed outline. Whatever form it takes, the purpose of the outline, from the writer's viewpoint, is to organize the research information into a logical, simple, natural progression.

It is for this reason and one other that the producer may want the writer to submit a formal content outline before proceeding to the treatment. The second reason is to gain client approval and be sure the project is still on track.

Simply put, then, the content outline is a factual document that *proves* to the producer and the client that the writer:

1. Understands the research material she has been studying.
2. Is including and *excluding* the proper facts in explaining the material.

3. Has found a logical order and flow that will make the material easily understood by the intended audience.

As you can see, the content outline does not deal with visualization at all. That is left to the treatment.

SHOULD EVERY PROJECT INCLUDE A CONTENT OUTLINE?

Like the different styles of scripting, shooting, and editing, everything has its place and its purpose in the production process. We have just discussed the purpose of the content outline. As for its place, consider two types of projects and two types of writers.

The first project is a strictly motivational program. Its purpose is to congratulate employees for a great job last year and encourage them to keep up the good work. The writer is someone you have dealt with many times, who you know can nail down exactly what you're after. The content is also very light, consisting mainly of a simple list of accomplishments for the previous year.

The second project is a long, detailed explanation of a piece of electronic gear, with a demonstration of how it should be used on the job. The writer on this project is new to you but comes with a very high recommendation from a fellow producer. In this case, the content will be extracted primarily from a thick user's manual, which the writer must learn from cover to cover.

Obviously, if either program should have a content outline developed, it's the second one, simply because the material on which it is based is much more complex and therefore much trickier to understand and present to others. In addition, although the writer on the second project may be highly recommended, he is still an unknown element to you and your client.

Subject matter complexity and the writer's experience are not the only two considerations that dictate when development of a content outline is appropriate, but they are two very good ones. Some corporate television departments require content outlines from their writers on every project; others do not require them at all. My suggestion is to let the content outline serve its purpose for you when it is appropriate.

I have included a partial content outline on the following pages, which should illustrate the appearance of the outline.

FIVE-STEP SAFETY:

Reducing the Probabilities

A Partial Content Outline

I. ACCIDENT PREVENTION--A MATTER OF PROBABILITIES

Accidents are bound to occur in virtually any occupation--especially one in which large numbers of employees work in an assortment of jobs and environments. Just as the number of those accidents in relation to the number of employees and hours worked is <u>predictable</u>, a specific reduction of accidents is also predictable.

The idea of influencing probabilities can be demonstrated with a pair of dice. Statistics show that in every 1,000 rolls of the dice, there is a high--and very predictable--probability that the number 7 will appear a certain number of times. The dice, however, can be ''loaded,'' thus influencing the probabilities of the appearance of 7s. In the same way, a management team can ''load'' employee attitudes, work habits, and environment, thereby influencing the number of times accidents should occur.

II. REDUCING PROBABILITIES--A SUPERVISORY RESPONSIBILITY

By law, employees are responsible to act in a safe manner on the job. Providing safety policies and procedures (the basis for those actions) is the responsibility of the employer. The application and enforcement of safety policies is, for the most part, the responsibility of the immediate supervisor. The supervisor can choose to handle this responsibility in a number of ways, each of which will have an influence on accident probabilities. As an example, if the responsibility is ignored, the probability of accidents increases. If applied after the fact, it might be effective to a small degree. At this point, however, damage has already been done that no amount of preventive medicine can change. If the responsibility is met in advance, the probability of accidents occurring should decrease.

This points up the need for the supervisor and all middle-management employees to develop action plans in advance that are specifically aimed at reducing accident probabilities.

III. THE SUPERVISOR'S ATTITUDE--A KEY

To be successful at influencing probabilities in advance, supervisors must first develop certain attitudes toward safety in general. Safety should be viewed, for instance, not as extra work but, rather, as one of a supervisor's many professional responsibilities. He or she should also keep in mind the fact that good safety records don't always mean that poor safety habits are not a problem in the work group. In many cases, it is simply a matter of time before an individual's or a group's poor safety habits catch up with them and accidents begin to occur.

The training and development of the people in the work group are also the responsibility of the supervisor. He or she should remember that a group is constantly being trained (in a positive or negative manner) by the attitudes to which its members are exposed. The issue, then, becomes a matter of the type of training and development the supervisor wants to achieve. A positive, action-oriented attitude toward safety on his or her part will develop similar attitudes in subordinate employees.

Finally, the supervisor should view the safety of all employees as an issue requiring the active involvement and open communication of everyone in the group.

IV. FIVE-STEP SAFETY--A PROBABILITY REDUCTION TOOL

The Five-Step Safety Program is an action-oriented tool for influencing accident probabilities. Although each step deals with a separate topic, all are highly interrelated and thus dependent on each other for success. The five steps are:

1. Education: This is the initial step in implementing any new program--creating an awareness of its objectives and how they will be achieved. Five-step safety education should take place in three ways:
   a. Formal training in company courses
   b. Ongoing dialogues between supervisory and craft employees during safety meetings
   c. Ongoing dialogues on a day-to-day basis

THE TREATMENT

The *treatment* is a first visualization of how the project will actually appear on the screen. It can be developed in several ways.

On the basis of the design factors brought out in the PNA, you may have a precise idea of exactly what you want the script to look like. In this case, you can simply tell the writer how you would like her to approach the subject.

More often, however, you probably will not have specific visual parameters. In that case, it is usually best at least to give the writer a set of general guidelines. You may, for instance, not be able to afford a full cast of actors. You may want the production to be done in a studio as opposed to on location. You may feel that humor would be a good format for this particular program, or that it would not.

These feelings are, again, something you cannot learn from a book. Developing a feel for what works comes with time, hard work, and a great many PNA reviews. But once you *do* develop a knack for knowing what's right, go with your hunches regardless of the pressures you may receive from writers to do it their way. Your instincts will usually turn out to be amazingly accurate.

The treatment should be a simple story narrative. Although some inclusion of camera directions, such as "close up" or "pan," may be needed to describe accurately how the scene will play, they should be kept to a minimum. The treatment should be kept uncluttered with script terminology the client may not understand, such as B.G. (background), F.G. (foreground), RACK FOCUS (change of focus from foreground to background or vice versa), and the like. Above all, it should provide a simple description of the program that does three very important things:

1. Describes visually what the program will look like on the screen.
2. Incorporates the content points that support the objectives developed in the PNA.
3. Allows the producer to estimate the resources needed—time, people, and money—to produce the program.

On the following pages, I have included a partial treatment example for a program on quality control.

QUALITY CONCEPTS

A Partial Program Treatment

We open with upbeat music and a distant shot of a car coming toward the camera on a freeway. The shot is from an overpass. As the car gets closer, we begin to see that it is large and silver in color. As it gets still closer, we are able to make out that it is a Rolls Royce. As it is just passing below us, the image freezes and the title is superimposed: ''Quality: What's the Price?''

We now go to a shot looking down a busy street (Twenty-second, facing south) and see that the same Rolls is just turning the corner some distance from us. It turns in our direction and approaches.

As the Rolls is about to reach us, it turns left. The camera pans with it, and the angle now reveals that we are just outside the Twenty-second Street Mall. The Rolls has pulled into the public parking lot.

As it pulls into a space, we see that the host of our program, a middle-aged first-line supervisor-type, is driving. He steps out and begins by using the car as a reference:

''I don't actually own this Rolls Royce. I couldn't afford it on a supervisor's pay. But I sure wish I could. It's definitely a top-quality vehicle.''

The host goes on to ask us what we think makes the Rolls Royce a quality vehicle--its price? Size? Amenities? Reputation? After giving us a moment to think it over, he answers the question by saying it's actually none of these things. Quality, he says, simply means meeting two strict sets of standards: the <u>technical</u> standards of a product's design and the <u>expectation</u> standards of the customers in the market for it.

''As an example,'' he says, ''let's compare the Rolls, here, in terms of quality, to a product of ours--a communications system.''

As he says this, we see a series of shots that compare parts of the Rolls to parts of one of our Compra Digital Systems.

On a shot of the Rolls interior, for instance, the host tells us that standards for the upholstery state that it must be steer leather of a certain kind and thickness and must be treated by a standard method to give it its soft, luxurious feel. As we now see the interior of a Compra Digital, the host tells us that the same high technical standards are present, but in relation to the types of chips used, the soldering methods, the spacing and layering of circuits, and so on.

We now see a shot of the exterior of the Rolls in which the deep silver paint is featured. Our host continues, ''And how about the outside?'' On the Rolls, it's a specific brand of paint, sprayed on to a standard thickness under controlled conditions and with particular types of equipment. The drying time is also per a standard. ''And for Compra Digital,'' he tells us, ''it's the same thing: The plastic cover is one of four standard colors and it's created under specific technical standards to give it depth, resilience, and shine.''

After another comparision--perhaps the Rolls's engine and the Compra's key pad--the host reiterates his point: One sign of quality is ''strict to the technical standards of the design of the product.''

We now find that the host has left the parking lot and is inside the mall, just outside a retail outlet that carries the Compra Digital.

He steps up to the store window, stops, and tells us: ''But remember, technical standards are only half the picture. The other half was the expectation standards of the customer. And what exactly are those expectations? That's what we're here to find out.'' As he says this, the host turns and enters the store.

We now see a series of three to five interview testimonials in which customers inside the store give our host their quality expectations about communications systems like the Compra Digital. Their comments include such descriptions as ''sturdy,'' ''fast,'' ''user-friendly,'' and ''covered by a satisfaction guarantee.''

A FINAL THOUGHT

One final thought on treatments: They do not always have to be in written form. It is perfectly acceptable to ask a writer to come back or even call with two or three verbal treatments that she can pitch to you. In this case, she avoids doing a lot of work that may not be what you want, and you get the benefit of being able to pick from several ideas. Following this verbal presentation, of course, she would write up the one you want and get it back to you. You and she would then present it to the client for discussion and approval.

Following approval of the treatment, you're ready to move on to the script.

4 The Script: Framework for Production

THE SCRIPT IN GENERAL

Corporate scripts can take different forms, and there are conflicting theories about what works best. Some producers say that a script should be very entertaining in order to hold audience attention. Others insist that although entertainment is acceptable, a script should be very tightly structured, with heavy use of learning points. Still others feel that a corporate script should not be entertaining at all because it is strictly an educational tool.

All three opinions are probably right at different times, in different situations. In other words, *the right type of script for the job is the one that best accomplishes that particular program's objectives, in keeping with the design factors brought out in the PNA.*

Visual Formats

Corporate programs also tend to fall into certain visual formats. When a program is heavily instructional, for instance, the use of learning points is essential. Often, this type of script calls for an on-camera host, who introduces the subject and walks and talks the viewer through it. At various times, cut-away footage is used to depict the points the host makes. Superimposed titles can act as an additional reinforcement of the message.

When a program is more demonstrative in nature or has to get across a concept or an idea, role-play is usually a good format. Although it may not communicate detailed information as well as an on-camera host, it leaves more room for exploring attitudes, ideas, and human subtleties.

Another widely used script format is one employing the voice-over narrator, in which the host is never seen but simply talks us through the subject while we see footage illustrating the points he makes.

Mixing formats is also widely accepted. For instance, a role-play scene may introduce the subject, perhaps in the form of a dilemma.

The on-camera host then enters the scene in the foreground and tells us he has the solution to this dilemma if we'll just pay attention. He then leads us through the material we are to learn. The role-play may recur periodically during the program to provide examples of points the host is making, or it may be seen again at the end as the dilemma is resolved.

Written Formats

Besides the visual format—host, role-play, and so on—there is also the script's *written* format to consider. Generally, there are two types of written script formats: *screenplay* style and *column* style.

The Screenplay-Style Script A script written in the screenplay style has screen and camera directions written across the entire page, and dialogue and narration typed in a column down the center. This is called a screenplay-style script because it is written in the same manner as feature film screenplays. It also lends itself to being broken down into segments for single-camera shooting—the way films are shot.

The Column Script The column-style script, by contrast, is written in two columns. The *audio* or *sound* column usually takes up the right-hand side of the page. This includes the narration, dialogue, sound effects, and music. The *video* or *picture* column takes up the left-hand side of the page and is made up primarily of the scene descriptions and camera directions.

The television or column-style script was developed primarily for multicamera television shooting. That is, it makes it easy for the television director to break the script down into the many "take" points that she will use to switch cameras live as the performance is underway.

Although the column style is probably more prevalent in the corporate world, the written format you choose should depend on the situation and on your personal preference.

BASIC SCRIPT ELEMENTS

Whatever the structure or format, all good scripts have several key elements in common:

1. They use the visual and sound media wisely—they are interesting to watch, and their dialogue or narration is written simply and naturally.
2. They are structured to flow naturally from one thought to another, with appropriate transitions.
3. They are simple, using any special effects economically and only when those effects help accomplish the program's purpose.

case study
Stick to Your Guns

One of the first projects I produced got off to a rocky start because I lacked confidence in my own judgment.

The project consisted of two scripts, which were parts 1 and 2 on the same subject—safe evacuation procedures. When I first read the scripts, I felt uneasy with them at once. They seemed wordy, rambling, and generally unfocused. The transitions seemed rough, and at times the writer changed tenses and interchanged the use of singular and plural.

At first, I was hesitant to say this to the writer because he had been in the business for years and I was a newcomer. Shouldn't he know more than I did about good writing?

After a good deal of mental wrestling, however, I decided that, newcomer or not, I was still the producer and was obligated to express my opinions. I would simply have the scripts revised. Determined to accomplish this, I called the writer in the next day.

My worst fears were immediately confirmed. Soon after I began to talk, he looked at me as if he couldn't quite believe that I, a brand-new producer, could possibly be challenging his writing ability. Moments later he went on to say so, civilly but with obvious distress.

To make a long story short, I was quickly intimidated and gave in, all the time trying to convince myself that since he was the veteran, he must be perfectly right, and that I just needed to learn more about the business.

As it turned out, however, he was not right. When the clients and my boss saw the scripts, their reactions were the same:

"Doesn't this beat around the bush a lot?"

"Does he mean singular or plural?"

"How about some transitions to smooth out the flow of this part?"

"These scripts really need work."

The lesson is a simple one. If you are sitting in the producer's chair, it's for a reason. Somebody trusts your judgment. Although there's nothing wrong with getting opinions and input from writers, directors, and others, in the end you are the one responsible for the project. Trust your own judgment. Decide what you want, and then stick to your guns until you get it.

---

4. They can be produced on a reasonable budget—for example, they are not filled with scenes it would take a feature film crew in Grand Central Station to record.

As with the treatment, I have included partial script samples on the following pages. In this case, however, there are two samples. The first is a role-play script written in the screenplay format. The second is an example of a column-style script with an on-camera hostess.

US AND THEM, COACH:

Closing the Management Gap

A Partial Screenplay-Style Script

FADE IN:

EXT. GTA HEADQUARTERS BUILDING--DAY

LONG SHOT, SLOW ZOOM IN. MUSIC UP AND SUPER TITLE: ''Us and Them, Coach: Closing the Management Gap!! As we LOSE TITLE AND MUSIC GOES UNDER, we hear the VOICE OF PAT MARSHALL, a newly promoted supervisor.

PAT (V.O.)
. . . Yes, Dell, I do realize productivity is falling off . . .

DISSOLVE TO:

INT. PAT'S OFFICE

CAMERA STARTS ON A CLOSE-UP of a name plate reading ''Pat Marshall, Supervisor, Customer Service.'' PANNING across the name plate, CAMERA FINDS a bottle of aspirin and a glass of water.

PAT (V.O. cont.)
Yes . . . I am working on a plan . . . As a matter of fact, I have some changes to put into effect on Monday.

CAMERA NOW FINDS AND HOLDS ON a certificate of achievement given to Pat as a recent graduate of Orientation to Management training for new supervisors. CAMERA finally PULLS OUT TO REVEAL Pat, looking very stressed and nervous, sitting at her desk with the phone to her ear.

PAT (cont.)
No . . . I . . . Well, I . . . I have to be frank with you, Dell. Morale is not very good . . . Yes, I'm working on that, too . . .

CAMERA RACK FOCUSES past Pat to TWO of her EMPLOYEES having a conversation in the B.G. It's obvious they're unhappy, and, as Pat says her next line, they shake their heads and gesture into her office as if talking about how incompetent she is.

PAT (cont.)
I agree. I think they just need to get used to me . . . and me to them. I'm sure we'll get along fine very soon.

CLOSE-UP--PAT

PAT (cont.)
Right . . . I'll keep you informed. Thank you.

As Pat hangs up, she lets out a deep breath. Her expression tells us how depressed and upset she is. She then picks up a memo from her desk.

ANOTHER ANGLE--OVER PAT'S SHOULDER--MEMO

The memo invites all customer service employees to the department's picnic and management-versus-craft softball game. AS CAMERA STARTS A SLOW ZOOM IN to the memo, we hear Pat say to herself . . .

PAT
(Shaking head)
Softball . . . All I need is another strike out . . . !

RACK FOCUS TO:

EXT. PARK--DAY

Beginning with a CLOSE-UP of a hand-painted sign attached to a bush, CAMERA HOLDS. The sign reads: ''GTA Customer Relations Employee Picnic.'' This is written over a drawing of a softball, bat, and mitt. CAMERA MOVES LEFT around the bush REVEALING a large group of people in the distance. They are gathered around picnic tables under a large tree. We HEAR laughter mixed with conversation.

TWO SHOT--FRED RANDALL, Pat's former supervisor, and Pat. They are seated at one of the picnic tables spread with food. We discover them in mid-conversation, between sips of coffee.

FRED
. . . and the potato salad is Helen's. Trust me. You'll love it . . .

Pat looks depressed.

PAT
It looks great . . . Does she still bring in her cookies and cakes on Fridays?

Fred pauses for a moment, reading the troubled tone in Pat's voice.

FRED
''Still''?! Hey, c'mon . . . It's only been three months since you got promoted. You talk like it's been years!

Pat glances down for a long moment at the coffee she's been sipping. Finally, she looks up. It's obvious Fred has just struck a nerve.

PAT
In some ways, Fred . . . It seems like years

FRED
Problems adjusting?

Pat acknowledges with a depressed nod.

FRED (cont.)
Your people giving you a hard time?

PAT
Not on purpose, I don't think . . . It's just . . . Well, it seems like they really don't understand why I'm there.

FRED
How so?

PAT
I honestly get the feeling they think every decision I make is either plain old stupid or just meant to give them a hard time! They don't realize I've got a job to do! (Pause) I'll be honest with you, Fred. I'm thinking about quitting this management business and going back to something I'm good at!

Fred eyes Pat for a moment with genuine concern. Then he smiles, suggesting he's figured out what her dilemma is.

FRED
Awe c'mon . . . Give up a perfectly good job over the old Management Gap . . . ?

PAT
Management Gap . . . ?

FRED
Yeah . . . It's very common. Basically an ''us and them'' attitude, right?

PAT

Right . . . ! I guess I'm a little naive in some departments . . . How about passing on some of your--

Another EMPLOYEE now enters with a softball, mitt, and bat over his shoulder. He cuts Pat off.

EMPLOYEE

You two wanna play a little ball before lunch or are you gonna talk business all morning?

Fred glances at Pat. It's obvious she's not in a ball-playing mood, and wants to hear more about what Fred was about to say. He turns back to the employee.

FRED

(Smiling)

You mean you want me on your team, Al?

EMPLOYEE

Well . . . Now I didn't say that . . .

After a chuckle and another glance at Pat, Fred turns back to the employee.

FRED

Tell you what . . . As much as I know it's gonna kill you guys to not have me in there, go ahead and pick teams . . . Pat and I may join you later . . .

The employee throws his hands up, and, as he exits, says . . .

EMPLOYEE

(Joking)

We'll try real hard to make it without you . . .

DISSOLVE TO:

EXT. PARK--SOFTBALL DIAMOND--DAY

The softball game is now in full swing. CAMERA STARTS on a LONG CLOSE-UP of the pitcher. She pitches, the ball is hit, and runners start around the bases to a round of cheers. CAMERA PANS, finding Fred and Pat in a high corner of a most unoccupied bleacher. Again, we have caught them in mid-sentence. Their AUDIO begins over the LONG SHOT of the game and continues as CAMERA ZOOMS IN.

FRED (V.O.)
. . . and it all boils down to a difference in perspective.

PAT (V.O.)
Sorry, but I don't quite follow . . . or agree. I think we should all have one perspective, and that's to do the best job we can.

FRED (V.O.)
No . . . not one perspective . . . one goal. And knowing the group you've got working for you, I'd say that's probably the case . . . They're all good people.

DISSOLVE TO:

TWO SHOT--Fred and Pat in bleachers.

PAT
Come on, Fred . . . Goals . . . Perspectives . . . You're talking semantics!

FRED
No I'm not . . . !

For a moment, Fred mentally searches for an example. Suddenly, we see he has it. NOTE: THROUGHOUT THE FOLLOWING DIALOGUE, AS PAT AND FRED DISCUSS VARIOUS PLAYERS AND POSITIONS, WE INTERCUT FOOTAGE OF THE GAME AS APPROPRIATE, AND AS NOTED.

FRED (cont.)
There's an example right down there . . .

WIDE SHOT of the game in progress.

PAT
The game?

FRED

Right . . . Take either of those teams. Let's say Alicia's. It's made up of players and a coach, right?

PAT

Sure.

CUSTOMER SATISFACTION: How It's Done

A Partial Column-Style Script

FADE IN: (MUSIC UP)

MONTAGE–A fast-paced series of SHOTS of employees on the phone with customers. We see them taking orders, researching accounts, and so on. We also see an assortment of expressions, which tell us some of the calls are going well and some not so well. When we reach an employee obviously having a bad time with a customer, we . . .

FREEZE FRAME

SUPER MAIN TITLE

Customer Satisfaction:
How It's Done!

DISSOLVE TO

INT. STUDIO SET

This is a stylized office set in black limbo. Our HOSTESS is seated at a Customer Rep's desk on which is a computer terminal. Several graphics are ''flown'' behind her--a blown-up Service Order, a Repair Order, and a Customer Account Research form.

(MUSIC UNDER AND OUT)

The hostess looks up, sees we are watching, and speaks. As she does so, she gets up from the desk, moves around, and leans against the front of it.

HOSTESS

You know, customer satisfaction is a pretty common word these days. And we all know what it means--keeping customers happy with the services and products we provide. Actually achieving customer satisfaction isn't always a simple matter, though--even if our products and services are the greatest. Why? Because it often depends on personal interactions like phone calls--and, let's face it, some customers aren't the easiest to deal with. The fact is, some of them can be frustrating and even downright rude. As Customer Reps, I'm sure that's no surprise to you.

The hostess thinks for a moment, then turns to ANOTHER ANGLE.

So how do we handle these hard-to-satisfy customers in a way that doesn't ruin your day and possibly cause them to go elsewhere for their needs? That's what this program is all about. During the next few minutes we'll examine some proven techniques in professional customer service . . .

On the hostess's last sentence, CAMERA ZOOMS IN on the blow-up of the Customer Account Research form. When we have it in FULL FRAME, we . . .

DISSOLVE TO

FREEZE FRAME of an employee on the phone with a customer. Over this SHOT we SUPER TITLES AS NOTED BELOW:

HOSTESS (V.O.)

Things like expressing genuine concern for the customer's feelings . . . making every effort to accomplish what the customer wants . . . and letting angry customers vent their frustrations without interruptions. . . . But let's start with that first one: expressing genuine concern.

GENUINE CONCERN

ACCOMPLISHING WHAT THE CUSTOMER WANTS

LETTING ANGRY CUSTOMERS VENT FRUSTRATIONS WITHOUT INTERRUPTION

We now LOSE ALL THREE TITLES, and our PICTURE UNFREEZES. It is a phone conversation between JENNIFER ANDERSON, a Customer Rep, and an angry female customer.

(JENNIFER'S AUDIO UNDER)

HOSTESS (V.O. CONT.)

And why not let an expert Customer Rep like Jennifer Anderson show us how it's done . . .

(JENNIFER'S AUDIO UP)

JENNIFER

. . . and you say it's been wrong for two months now?

CUSTOMER'S VOICE

Not only wrong . . . it's been off by over twenty dollars both months!

JENNIFER

One moment, Mrs. Neal. Let me look up your records.

Jennifer keys several commands into her computer terminal.

CLOSE-UP--COMPUTER SCREEN.
We see the account of Alice Neal appear.

BACK TO JENNIFER

JENNIFER

I have your account, Mrs. Neal, and I do see there's been a correction both times. I'm really sorry about this. I can see how it's probably been upsetting for you.

CUSTOMER'S VOICE

It's been maddening!

THE FIRST DRAFT

When the first draft of the script is submitted, you may want to send it to the client for immediate input, or you may prefer to meet with the writer and have a second draft written first. The latter is the better of the two options unless you feel the first draft is very close to what you want.

Eventually, however, it will go to the client, and she will provide input. If you are lucky, you will get the input of one or two key decision makers (your client and perhaps her boss or a content expert) who are willing to comment on the content only and leave the media decisions to you. If you are unlucky, you will receive the comments of a committee or group of people, each with different ideas and only one thing in common—each feels that his input is indispensable.

In the good scenario, you'll have no problem. You simply consider the input, apply your skills in media development to the content the client wants to include or exclude, and then have the writer produce a new draft based on what you feel is right for the project.

In the unlucky scenario, chances are you're in for *big* problems. It is common knowledge in the corporate media business that making programs by committee is usually disastrous.

Why? If each person on this committee feels that his or her ideas are indispensable, and if you try to accommodate all of them, you may just spend the rest of your corporate television career attending script meetings, writing checks to the writer, and trying to explain to your boss why this particular program is going nowhere.

The solution is simple—sometimes. Let your client know from the outset that in order to produce her program in a high-quality, cost-effective way, you need her complete trust in making the aesthetic decisions. Also let her know you will need a *single* key decision maker when it comes to any other input that might arise during the many phases of production.

Inform her that if she wants to get input from other departments or individuals, as is often the case, this is fine, but ask that she take those comments as suggestions only and, whenever possible, give you the final say as to whether or not the information should really go in the script. Explain that without this client trust and weeding out of extensive comments, your job can become extremely difficult or even impossible.

In some corporations, this last scenario is wishful thinking. In the end, you *will* have to face a committee no matter what you say or do. In such a case, I can only wish you the best of luck and advise you to hang in there until you finally arrive at a completed, approved shooting script and the beginning of preproduction.

case study
Approval by Committee?

I was once asked to develop a program that would teach employees the principles of high-quality workmanship.

In the initial client meeting, when I asked who would be providing the approvals at each step during the process, my client assured me that he would be the one. His boss, he said, had delegated the project to him and had given him complete authority to make all decisions about its development.

I had heard this before from clients and later found that, in the end, the boss really did want some sort of input into the project. So I suggested to my client that even though he had the final say, it probably wouldn't hurt at least to keep his boss informed as we progressed. After some discussion, he agreed, and we moved forward on the project.

On the basis of the audience analysis information contained in the PNA, I suggested to the client that a role-play format would work well for the program. I reasoned that a well-done role-play would allow us to demonstrate the principles of quality as it related to the jobs of audience members and, at the same time, let us explore the general concept of what *quality* means to any business.

My client agreed wholeheartedly, and the program went through script approval and into production with his approval at every step of the way. It was only when I had completed the off-line edit and called him in to view it that the trouble began. After watching the show, he said he liked it very much and announced that he would like to take a copy to the Quality Steering *Committee* for their input.

Tactfully, I asked who was on this committee and why he wanted them to have input on the program now when they hadn't been involved in the proposal, treatment, script, and so on. Realizing that I was upset, he assured me there would be no problem: They would love the program as much as he did. Showing it to them, he insisted, was simply a formality.

How wrong he was!

The vice-president who headed the committee had his own ideas about what a program on quality should look like, and those ideas didn't include role-play. What he wanted instead was a documentary program—interviews with quality control managers, department heads, supervisors, and employees on the line. He turned out to be a very forceful person who saw things only one way—his way.

I attended the next committee meeting to defend the approach we had taken. I told the vice-president and the rest of the committee that the program had been designed on the basis of audience analysis, and I assured them that the approach we had chosen would be very effective. To make my point, I passed out copies of the approved PNA and led the committee through the audience analysis section.

Although most committee members seemed to agree with me, it soon became obvious that none of them—*including* my client with "full approval authority"—was about to buck the VP's wishes. And no matter what I said, he wanted a *documentary*—period.

The result? As usual, what this vice-president wanted, he got. The first program was shelved, and a new program—a documentary—was produced to replace it.

There are two lessons to be learned from this story. First, be sure to get the input of the person or persons who will actually approve your program *as you develop it*—not once it's nearly finished. Second, there are times when, despite your expertise and all your attempts to do a project right, you will simply be overruled because of politics. When this happens, you must be professional enough to realize that at a certain point, more resistance can only cause personal or departmental damage. When you reach this point, your only choice is to accept the situation and do what you have to do gracefully.

part two

Preproduction

5 The Plan for Success

DETAIL—THE KEY

Preproduction involves the organization, confirmation, and scheduling of all elements of production. It is a critically important prerequisite to a smooth, productive shoot. That's because preproduction is the time for attending to details. And if *anything* will stop a shoot in its tracks, it's some silly, perfectly obvious detail no one bothered to think about in advance.

For example, I once directed a corporate music video that required shooting employees running through the woods at night. We had no real woods close by, but I found a local golf course that, if lit and shot properly, would double nicely for woods.

We made all the arrangements and attended to what seemed to be everything in advance: Who would open up? Where was power available? Could good-quality sound be recorded? At what times did golf course employees leave and arrive? Were there any restrictions? What was the phone number of the person in charge?

We arrived at the course at dusk ready for a full night of production. The crew began unloading the vehicles, and I got out my script book and walked with the director of photography to the first camera location. As we arrived under a stand of oak trees and began discussing the shot, we both suddenly heard a loud hissing sound. We swung around in our tracks to find that the sprinklers—huge industrial, rotating rainbird types—had come on and were about to soak us.

It turned out that the sprinklers were all on timers, preset to come on periodically in different places throughout the night. Fortunately, we were able to reach the superintendent in charge of the grounds, who turned them off for us. Had we not been able to reach her, however, the shoot would have been ruined. And had the sprinklers not come on until after we were in the trees actually shooting, not only would some very expensive equipment probably have been ruined, but the electricity being used to power the lights would have been an extreme safety hazard to everyone close by.

THE PROCESS

The preproduction process itself usually involves four key people: the client, the producer, the director, and an assistant director. In some cases, a production assistant is also brought in to help.

The client's role is primarily one of support. She works closely with the asistant director or production assistant in arranging for things like locations, props, and employees who may be in the program. Often, the client will also act as a liaison between the production team and various executives or departments in the company. She may also approve certain aspects of preproduction—talent, locations, shooting schedule, and so on.

Generally speaking, the producer's job during preproduction is to budget the production and to oversee, guide, and approve the activities and elements being brought together by everyone else involved. This means approving or disapproving locations, wardrobe, props, talent, and the like. It also means approving the director's shooting schedule and planned approach in terms of visualization, coverage, crew size, equipment needed, and so forth.

The director, after reading and visualizing the script, decides what elements and activities are needed and obtainable within the established budget and the producer's guidelines. The director also plays a key role in casting and establishes the basic studio or location plan—a shooting schedule and shot list—to accomplish the production.

The assistant director works with the director in making all the production arrangements. A good assistant director will take care of virtually all the details involved in bringing the production on line, while allowing the director time to work primarily with her script, casting, and locations. In addition, a good assistant director assumes *nothing;* he confirms and double-checks *everything.*

If a production assistant is brought in, he usually works for the assistant director attending to typing, deliveries, certain phone calls, and other details.

Costs

As mentioned previously, corporate budgets do not always allow for a full staff of people to carry out the preproduction process. Assuming yours does, the following are ballpark cost figures for the preproduction team just discussed:

Position	*Cost per Day*
Director	\$300—\$500
Assistant director	\$100—\$300
Production assistant	\$ 50—\$150

Unless otherwise negotiated, these fees are usually payable in a lump sum after the production is completed.

Preproduction Tasks

What exactly will these fees buy for you? The following is a general list of the tasks this team will accomplish during a typical preproduction.

1. Budgeting
2. Reviewing and breaking down the script
3. Obtaining props and wardrobe
4. Scouting and confirming locations and obtaining permits to shoot
5. Auditioning and selecting talent, professional and otherwise
6. Writing the shooting schedule
7. Hiring the crew
8. Building and lighting sets
9. Renting or reserving production equipment and vehicles
10. Designing and creating artwork and any character-generated titles
11. Blocking the script
12. Rehearsing
13. Conducting preproduction meetings
14. Preparing equipment for the shoot

Obviously, preproduction is a very demanding part of the overall production process. As I've mentioned, what happens during this time can literally make or break the production days themselves. With that in mind, let's cover each of these preproduction elements in greater detail.

6 The Elements of Preproduction

BUDGETING

There are a variety of opinions on when a program should be budgeted. Some producers feel that budgeting should occur at the treatment stage, because this is when the visual foundation of the program is worked out. Others feel budgeting should be done even earlier by establishing a sum of money at the outset of a project and making the project fit that monetary mold.

Although this latter form of budgeting is an excellent cost-control measure, it can sometimes make programs less effective because they may be designed on the basis of costs rather than such PNA factors as audience, objectives, longevity, or potential monetary impact on the company.

Budgeting is probably most effective at the beginning of the preproduction stage. Only with a shooting script in hand can a producer accurately forecast the cost of a production. Establishing a preliminary budget *range* following the proposal stage, however, has two benefits. It allows for a certain amount of cost control up front before the client's basic needs have been worked out, and it allows for increases or decreases as the proper design begins to take shape.

Assuming that budgeting is accomplished at the preproduction stage, the producer budgets a program by breaking it down into basic elements: How many actors are needed, and for how long? How much and what type of equipment is needed, and for how long? What will the crew size be? How much did the script cost? How much and what type of postproduction is needed? Figure 6.1 is an example of a typical production budget.

REVIEWING AND BREAKING DOWN THE SCRIPT

Just as the producer reviews and breaks down the script, so does the director, but for different reasons. The director must know every detail

Project # 48-D Date 7/8 Title "The supervisor...."

Producer Dan G.. Director Martin

Writing/Design	Qty	Rate	Days/#	Total
Producer	1	250	2	500
Writer	1	1800	(FLAT)	1800
Rewrites				
Supp. Materials				
Messenger/Post	1	2	—	8.—
Other 1				
Other 2				
		TOTAL W & D		2308

Pre-Production	Qty	Rate	Days/#	Total
Producer	1	250	2	500
Director	1	400	4	1600
AD	1	200	4	800
PA	1	100	1	100
Dir Of Photo	1	250	5	125
Cam				
Engineer				
Audio				
Teleprompter				
Gaffer				
Grip				

Figure 6.1 *Production budget. Most production budgets are designed to list all line item expenses involved in the program development process. Subtotals for the major sections—writing and design, preproduction, production, postproduction, and miscellaneous—are usually carried forward and combined to form a grand total. Line items listed as "other" account for unplanned expenses.*

	Qty	Rate	Days/#	Total
Flr Mgr				
Art Work	2	–	FLAT	90
Set Materials				
Props	–	–	FLAT	200
Travel				
Messenger/Post	^	–	FLAT	32
Rentals				
Other 1				
Other 2				
Other 3				
			TOTAL PRE PR.	3447

Production	Qty	Rate	Days/#	Total
Producer	1	250	1	250
Director	1	400	2	800
AD	1	200	2	400
PA				
Dir Of Photo	1	250	2	500
Cam				
Engineer	1	150	2	300
Audio				
Teleprompter	1	125	1	125
Gaffer	1	150	2	300
Grip				
Flr Mgr				
Props				

Figure 6.1 *(cont.)*

	Qty	Rate	Days/#	Total
Travel/lodging				
Meals	8	7.50	2	120
Rentals	1	50	2	100
Talent	3	300	2	1800
Extras	2	50	2	200
Make-up				
Wardrobe				
Messenger/Post				
Studio				
Permits			FLAT	300
Other 1				
Other 2				
Other 3				
			TOTAL PROD.	5345

Post Production				
Producer	1	250	2	500
Director	1	400	1	400
Off-line Editor	1	200	5	1000
On-line Editor	1	250	2	500
Engineer				
Rentals				
Transfers/Dup.			FLAT	225
Music fees	2	80		160
Dist. Dup.	30	18		540
Messenger/Post	3	20		60

Figure 6.1 *(cont.)*

	Qty	Rate	Days/#	Total
Other 1 Aurora	1	100/HR	3 HR	300
Other 2				
Other 3				
			TOTAL POST PROD.	3685

Miscellaneous				
Video tape	15	18	—	270
Audio tape	5	7	—	35
Expendables	—	—	35	35
Telephone				
Xerox				
Insurance				
Other 1				
Other 2				
Other 3				
			TOTAL MISC.	340

TOTAL W & D	2308
TOTAL PRE PROD.	3447
TOTAL PROD.	5345
TOTAL POST PROD.	3685
TOTAL MISC.	340
GRAND TOTAL	15,125

Figure 6.1 *(cont.)*

of the script intimately in order to assess exactly how she will faithfully transfer it to videotape. For this reason, script review and breakdown are usually her first order of business.

Typically, a copy of the director's breakdown is also used by the assistant director, who immediately gets busy organizing details, booking people, typing contracts, finding props and locations, and scheduling people and equipment.

Although script breakdowns can take many forms, their basic purpose is to itemize and organize all elements of the script that will require some action in preproduction. The two examples in Figures 6.2 and 6.3 should serve as illustrations. Figure 6.2 is a breakdown by ele-

SCRIPT BREAKDOWN PROJECT # 8182 TITLE Safety...

TALENT

Host
(2) interviewees -
-Accident victim - Seat belt
-Accident victim - On-the-job
C.E.O.-Ins. Co. (If not C.E.O., an executive or mgr.)
Blk. or Hisp. worker - Grinding wheel
Female Executive
Secretary
Husband changing lightbulb on ladder
Blk. or Hisp. worker - Lifting
Man - Car passenger
Woman - Car passenger

LOCATIONS

Office Bldg. Ext.
(3) office Int.'s
- (Turn Cam.)
Office parking lot - or driveway
Shop - Grinding wheel
Shop - Stored boxes
Home - Hallway for lightbulb shot, or kit.
- Study - for interview
- Living room for interview or possibly kitchen

PROPS

(1) Business envelope
"A" frame ladder - 4'
(1) Lightbulb
(1) Pair goggles
(1) New/nice car w/ lap type seat belts
Several industrial boxes - one heavy
Various pieces of office furniture, dressing, plants, etc.

TITLES

Opening title
L/3rd ID titles
C.E.O
Both interviewees
Closing Logo

Figure 6.2 *Script breakdown #1. A simple script breakdown itemizes all elements contained in the script that need some attention. In corporate television, these are often limited to talent (actors), props, locations, and wardrobe. Breakdowns may also include artwork, special effects, character-generation requirements, and so on.*

PRODUCTION: # 71108-A
DIRECTOR: ARDEN

DATE: 4/11
PAGE: 1

SCRIPT BREAKDOWN

SCN.	MASTER DESCRIPTION	INT./ EXT.	PRINCIPLES	EXTRAS	SET-UPS	PAGE(S)	PROPS	SPECIAL EQUIP.	OTHER CONSIDERATIONS	SHOOT HRS.
7	HOST ENTERS @ DESK	INT.	HOST	2	1	1.5	—	—	CONTINUITY W/SCENE 6	1.0
12	JANET ON PHONE W/ SUPERVISOR	INT.	JANET	—	2	.5	CALENDAR NOTE PAD	DOLLY	—	1.25
12A	SUPERVISOR ON PHONE TO JANET	INT.	SUPERVISOR	1. FORKLIFT OPERATOR-B.G.	2	.75	FORKLIFT HARD HAT NOTEPAD	—		1.5
1	TRUCK DRIVES UP TO FACTORY HOST EXITS (L.S)	EXT.	HOST	2 IN PARKING LOT	1	.25	TRUCK	—		1.0
1A	M.S. HOST EXITS TRUCK AND ENTERS BLDG.	EXT.	HOST	—	1	-	—			.75
5	W.S. EMPLOYEES ARGUE IN PARKING LOT	EXT.	2 EMPLOYEES	2 IN LOT	2	2.75	CLIPBOARD HARD HATS	DOLLY		2.0
5A	EMPLOYEE SPEEDS OUT OF LOT	EXT.	2 EMPLOYEES	2	2	.25	CAR	—		1.0
5B	EMPLOYEE HOLLARS AS OTHER LEAVES	EXT.	2 EMPLOYEES	-	2	-	CAR	—		.5

(NOTE ITEMS TOO LENGTHY TO LIST ON ATTACHMENT OR REVERSE SIDE)

Figure 6.3 *Script breakdown #2. Directors sometimes use a scene breakdown such as this to estimate the time, number of setups, and special requirements of each scene in the script. This breakdown would begin in script order, as it is shown. The director might then cut the breakdown into horizontal strips and place them in an economical shooting order. This then becomes a first step in developing a shooting schedule.*

ments that would be developed and used by the assistant director as a basis for his line-up work. Figure 6.3 is a breakdown by scene, which would be developed by the director as a preliminary step in developing a shooting schedule.

OBTAINING PROPS AND WARDROBE

This task falls to the assistant director or the production assistant, often with assistance from the client. It can be as simple as obtaining a set of work tools for a scene in which a craft person is to appear, or as complex as visiting wardrobe rental facilities and getting costumes chosen, actors measured and fitted, and so on. In corporate television, the former is usually the case. But although obtaining props and wardrobe may seem a simple matter, it should never be treated as unimportant. A missing set of tools can hold up an entire crew for hours if the props were not properly arranged for in preproduction.

SCOUTING AND CONFIRMING LOCATIONS AND OBTAINING PERMITS TO SHOOT

In most cases, the director scouts locations early in preproduction, for a number of reasons. First, she needs to begin to visualize how a scene will play out in an environment for the camera and to be sure the location is visually representative of what the script is trying to convey.

Second, she needs to confirm that an environment that *looks* right will also work logistically. Can she park her crew's trucks nearby? Will she need a generator to provide power? Will she be able to get good-quality sound, or is heavy construction going on right next door? Will the sun be at the proper position for the time of day at which she plans to shoot? What kind of lighting equipment will be needed to compensate?

In corporate television production, another common consideration is how disruptive a production crew will be to the work going on in an employee location. Will the area have to be cleared? If a breaker is overloaded with lights and opens a circuit, will computers go down? Should the shooting take place during nonbusiness hours? If so, who will unlock the doors? Will someone need to move furniture? And so on.

Shooting on city streets or on private property necessitates permits and, usually, proof of liability insurance. In addition, depending on what is being shot and how, a police officer or fire officer may be required on the set. The assistant director or production assistant (PA) usually obtains permits from the appropriate city or county offices, or works with any private parties involved. Shooting without permits, though sometimes tempting, is risky because passing police officers can and often do shut down unregistered shoots on the spot.

Advance location scouting and confirmation are crucial to a smooth production. A wise producer considers them standard operating procedures on all shoots unless budget or circumstances absolutely do not allow for them.

AUDITIONING AND SELECTING TALENT, PROFESSIONAL AND OTHERWISE

A director I once knew used to say that directing is actually 90 percent casting and 10 percent luck. Although I won't give casting quite that level of importance, I will say that it, too, is crucial to a good final product. The reason is simple: Poor performances by actors or company employees can instantly ruin a program's credibility.

Preliminary casting is usually carried out by the assistant director. On the basis of what the director has requested in terms of actors—sex, age, type, and so on—the assistant director will normally work with casting books, pictures, other programs, or agencies. From these sources, she will find a number of choices for the director to review. From this material, the director then selects those actors she would like called in for auditions. A day and time is set up for the auditions to take place.

In many cases, the producer also attends the auditions, and in some cases the client is there, too. The danger of having too many people present is the same as with script approval: Casting can become a selection process by committee, which creates a very difficult situation for the director and, in some cases, is damaging to the program as well. Toward this end, I would advise keeping clients away from auditions if at all possible and simply hiring directors with a dependable eye for talent.

When it's company talent being selected, it is often not a true audition but, rather, an interview by the director that determines the selection. This would be the case, for example, if the program being produced were to include interviews with employees. The director would meet with the individuals to get a feel for how they will look on camera and how well they are able to express themselves verbally.

The use of company employees as actors in role-play scenes, with scripted lines, usually ends up being both embarrassing to the employee and a disaster for the program's credibility. My recommendation is simply not to do it.

Whatever the situation, using company talent in any capacity is always a risky proposition. Many company employees who are eloquent and expressive in conversation find it impossible to say ten words in a row once the lights and camera arrive. Although this sometimes cannot be helped, a good director knows she can save herself and the producer a lot of headaches by preinterviewing and carefully selecting the employees she will use.

WRITING THE SHOOTING SCHEDULE

This is another very important job of the director. On the basis of all the elements she has brought together and the number and complexity of her shots, she makes three determinations: (1) the best order in which to shoot the material, (2) how long each scene will take to shoot, and (3) the most economical way to put these elements together into an overall schedule.

As an example, consider a program that will be made up of three basic elements: a host being shot on location, employee interviews, and general footage of several work functions.

If the host will be featured at the same locations as the work functions, it may be best to visit that location only once and shoot both the host material and the work function. On the other hand, this may extend the number of days the host is needed, because he will be doing a lot of sitting around between his scenes. The question then becomes: Which will be more economical—having the host for an extra day, or shooting all his scenes in one day and then revisiting certain locations to shoot the work functions.

The availability of the employees being interviewed may also affect this decision, as will many other factors the director has had to deal with in preproduction.

Before it is finalized, the shooting schedule should be approved by the producer. The end result should be a shooting schedule that uses crew, talent, equipment, and locations as cost-effectively as possible. It will probably look something like the one shown in Figure 6.4.

HIRING THE CREW

Some directors like to pick their own crews, whereas others leave this to the assistant director. Typically, a director will be most interested in his director of photography (DP)—the person responsible for the picture quality of his program.

Crew sizes, like so many other factors in television production, will vary depending on the budget and the type and complexity of the shoot. Corporate shoots are often staffed with only three people: the producer,

SHOOTING SCHEDULE: SAFETY: THE CHEAPEST PREMIUM

DATE	DAY	TIME	ITEM
2/25	Mon.	12 NOON	Call time. Crew meeting & leave to pick up equip.
"	"	3:PM	Arrive @ John's. Set up for master—scn 190.
"	"	3:45	Shoot L.S. from across street, M.W.S from front of car. Low angle.
"	"	4:00	Set up for, and shoot O.S.
"	"	4:45	Set up for, and shoot O.S.and CU, her. Also, insert—belt being buckled.
"	"	5:30	Move inside, set up for first interview.
"	"	6:30	Shoot interview—scn. 200. (Robert D.) Lvng Rm.
"	"	6:45	Set up for second interview.
"	"	7:15	Shoot interview—scn 210.
"	"	7:30	Dinner. Pizza/Salad.
"	"	8:30	Set up for C.E.O. interview.
"	"	9:15	Shoot interview—scn 220.

Figure 6.4 *Shooting schedule. A shooting schedule can be broad in scope, listing only major parts of production activity, or it can be more detailed, as is the example shown. The director should provide a shooting schedule for the producer. It should be used to gauge, first, how economically the shoot has been set up, and, second, how efficiently and productively the shoot actually was carried off.*

DATE	DAY	TIME	ITEM
"	"	9:30	Set up in garage for "wife" on ladder-master.
"	"	10:30	Shoot master-scn 250. Patti.
"	"	10:45	Shoot CU and insert-feet and face.
"	"	11:15	Strike. Prep equipment for next day.
2/26	"	6:00	Call time. Leave for insurance co.
"	"	6:20	Arrive, set up for Host open & lock off of sign. David arrives, 7:00.
"	"	7:30	Shoot scn 100-Lock off of sign @ head of shot, then M.W.S. & CU Host.
"	"	8:00	Set up for second angle.
"	"	8:30	Shoot scn 110-MCU, ZOOM OUT, MS & CU.
"	"	9:00	Strike & wrap location. Head for Townsgate.
"	"	9:45	Arrive T.T.--break.
"	"	10:00	Unload & set up for Host title shot, w/employee.
"	"	11:00	Shoot scn 120-WS, ZOOM IN, MWS & CU.

Figure 6.4 *(cont.)*

who also doubles as the director; a director of photography, who sets her own lights and equipment; and a sound engineer, who also takes basic script notes or tape logs.

All of these crew members typically also carry cables and equipment and help set up and strike at each location. Normally, this type of crew would handle simple shoots such as event documentations, employee interviews, and general footage of employees at work.

A fully staffed, typical location shoot on a more complex production, including a host, specifically blocked scenes, and multiple distant locations would probably include the following:

Director
Assistant director/script supervisor
Director of photography/camera operator
Engineer in charge/sound
Gaffer
Grip
Teleprompter operator

A three-camera studio shoot, on the other hand, would include:

Director
Assistant director/script supervisor

Figure 6.5 *Small location crew. The director/camera operator rolls for a visual slate, which is held up by the grip. The sound engineer holds a shotgun microphone in place for the verbal slate. The mixer hung around the sound engineer's neck is wired back to a camcorder. The AD is down the street signaling a truck driver to proceed through the shot.*

Technical director
Director of photography/camera 1 operator
Camera 2 operator
Camera 3 operator
Sound engineer
Floor manager

Typical fees for these crew members (not including those people we have already discussed) would fall into these approximate ranges:

Position	*Cost per Day*
Director of photography	$150–$350
Camera operator	$100–$200
Technical director	$150–$250
Sound engineer	$150–$250
Gaffer	$100–$200
Grip	$75–$175
Floor manager	$75–$175
Teleprompter operator	$75–$150

Like the other fees we have talked about, payment to these crew members would normally be handled in one lump sum following the shoot.

We will take a closer look at what crew members actually do in Part III, on production.

In developing a stable of crew people, as is the norm, it is a good idea to start the same way as with writers—by interview and word of mouth with other production facilities.

BUILDING AND LIGHTING SETS

If the production is to take place in a studio, you'll need a set. This can be as simple as a stool and a flip chart or it can be much more complex. In any case, it will have to be designed, built, and lit prior to production.

Simple sets are typically done under the supervision of the director by the director of photography, one or two grips, and a gaffer. If the set is more complex, it may also require the help of an art director and of people who specialize in set design and construction.

Simple sets may take only a few hours to complete. More complex ones can take several days. Normally, this job is timed to be completed just prior to production, since the studio in which the set is built is usually being rented for several hundred dollars a day.

Figure 6.6 *Skeleton set. Studio with set walls erected. The crew must now bring in furniture, props, wall dressing, and so on.*

RENTING OR RESERVING PRODUCTION EQUIPMENT AND VEHICLES

Obviously, a production requires a camera, lights, a recorder, microphones, monitors, cables, and so on. If your company has its own equipment, the assistant director can simply reserve it for the period of the shoot. If the equipment is to be rented, she would call a rental facility and arrange to have it prepared and possibly delivered the afternoon prior to production. If delivery is impossible, a PA or crew member would have to be sent to pick up the equipment and, of course, return it after the shoot.

Most rental companies charge only for the production days but allow the equipment to be picked up in the afternoon of the day before and returned on the morning after the shoot. Many also offer reduced rates if the rental period is long enough. For example, if the equipment is to be rented for seven days, it is common to get a four- or five-day rate on it.

If we were to equip the simple three-person shoot mentioned earlier, the list would probably look like this:

Camera, power supply, and batteries
Head, tripod, and spreaders
Recorder, power supply, and batteries

(a)

(b)

Figure 6.7 *Three different types of sets: (a) An abstract limbo set for a host or hostess. This one is made up of a cluster of geometric set pieces, a single fern, and three pieces of lattice "flown" on monofilament line. (b) A common interview set. The program host (interviewer) would sit on the left in the single chair. Two interviewees would sit on the right. This setup works well for a multicamera format. Camera 1 would be placed to the left-hand side of this picture and would be used for* singles *of each*

(c)

interviewee. Camera 2, placed in the center, would be used for a wide three shot, *similar to this photograph. Camera 3 would be placed to the right-hand side of this picture and would provide a single of the host or hostess. A fourth camera might be used for art cards. (c) This bank interior set would be used for a role-play segment. The studio camera has been rolled in between two desks at center of picture. This was probably done to give the director an idea of what a certain angle might look like.*

Figure 6.8 *Clients on the set. Clients talk over the accuracy of the set design with the producer. Clients play a key role in handling the many details that arise in preproduction.*

Several microphones with batteries
Portable audio mixer with batteries
One light kit
Several rolls of videotape
Video cables
Audio cables
A/C cables
Gaffer's tape
Dulling spray

An equipment list for the more complex shoot would look something like this:

Camera, power supply, and batteries
Head, tripod, and spreaders
Recorder, power supply, and batteries
Video monitor with batteries
Waveform monitor with batteries
Several microphones with batteries
Portable audio mixer with batteries
Two light kits with gels/scrims/diffusion
Two reflectors
Several rolls of videotape
Video cables
Audio cables
A/C cables
Generator
Several C stands
Assorted flags and scrims
Gaffer's tape
Dulling spray
Makeup kit
Slate
Felt tip markers

As with the job functions of crew members, we will take a closer look at the purpose all this equipment serves in the section on production.

DESIGNING AND CREATING ARTWORK AND CHARACTER-GENERATED TITLES

Many corporate television programs need some form of artwork, usually diagrams, charts, or specially designed slides. This category also includes electronically generated titles, which need to be designed and in some cases made ready for production.

case study

The Value of the Preproduction Equipment Check

A major brush fire once burned through the operating territory of the communications company I was working for.

When I arrived at work that day, my boss approached me at once. The fire had just been contained, he said, and work crews from our company were being allowed in to begin rebuilding all the lines that had been destroyed. Since this had a direct impact on many employees who would never actually see the destruction in person, he asked that I immediately put together a small crew, roll out to the location, and see if I could produce a short documentary-type program for immediate release.

While an assistant began lining up a camera operator and sound engineer for me, I immediately got out the company phone book and began making calls. I soon found out where our crews were working and got the name of the administrator in charge of the restoration.

It suddenly struck me that a series of interviews with the person in charge as well as the people on the front lines would provide an excellent foundation for my program. Better yet, if I could get the top person out at the location among the crews and the rubble, my story would be sensational.

Getting interviews with the front-line employees would be no problem, since they were already there. But getting to the manager in charge turned out to be a different story. When I discussed the idea with her on the phone, she immediately told me there was no way she could leave her office. After I had pleaded my case for some time, explaining how important it was to other employees to see what was actually being done, she told me to call back in an hour and she would see what she could do.

Forty-five minutes later I called back, and she asked how long it would take. I told her we could meet at a predetermined location and have the interview done in fifteen minutes. Finally she conceded, although she made it clear that leaving her command post to one of her managers was a risky move on her part.

I assured her of my gratitude, hung up the phone, grabbed a note pad, and met with the crew at the rear of our van. My first question was: Do we have everything we need—camera, batteries, sound gear, videotape and so on? The answer was yes, and because we were in a rush I didn't bother to double-check.

When we reached the location, a mobile home park that had literally burned to the ground, I realized at once how well the interview would play. Crews were working frantically on all sides, homeowners were searching amid the rubble for any possession that might be saved, fire trucks were everywhere. To have the woman in charge of our restoration telling the company's story in the midst of all this would be perfect.

I chose a camera angle and briefed the crew, and they immediately began setting up. A few minutes later the administrator showed up, looking harried and impatient. I quickly assured her it would be only a few minutes and reiterated my appreciation that she had taken the time to come.

As we were discussing the interview questions I would be asking, my camera operator approached me sheepishly and asked if we could have a word. When we stepped to one side, he gave me the bad news: He had forgotten the batteries for the camera. There was no way we could shoot unless we could have someone deliver the batteries to us. The driving time was forty minutes.

Acutely embarrassed, I returned to the administrator, told her of our mistake, and asked if she could wait. She promptly turned me down, hopped in her car, and left.

From that day on, I made it an unbreakable rule: No matter what the circumstances, no crew of mine would ever leave the facility without an item-by-item equipment check.

In most cases, getting this accomplished is the job of the assistant director working under the director's supervision. It might be done through an in-house graphics department, on your own in-house character generator, or by a company outside of your own. Whatever the case, probably the most important thing to keep in mind in this area is lead time, since the design and production of artwork can sometimes take several weeks.

Designing art and character-generated titles also tends to be a part of the preproduction process that is left undone or is done with too little thought to how the material will actually be used in the program. The result can be a costly, time-consuming, and frustrating delay of the program until the material can be re-produced or repaired.

BLOCKING THE SCRIPT

This is the director's job. It involves preplanning how the actors will move in relation to the camera.

Some directors block a script by story-boarding it scene by scene, as shown in Figure 6.9. Others like to make overhead sketches showing the position and movement of the camera and the actors. Figure 6.10 provides an example. However it is accomplished, it's important that blocking be done primarily in preproduction.

I use the word *primarily* for two reasons. First, blocking tends to undergo fine tuning once the camera and actors are in position on location. The director may decide that the way the host was going to exit does not look quite right against a certain background. Or, if she feels a higher or lower angle is better, the action may have to change to some degree.

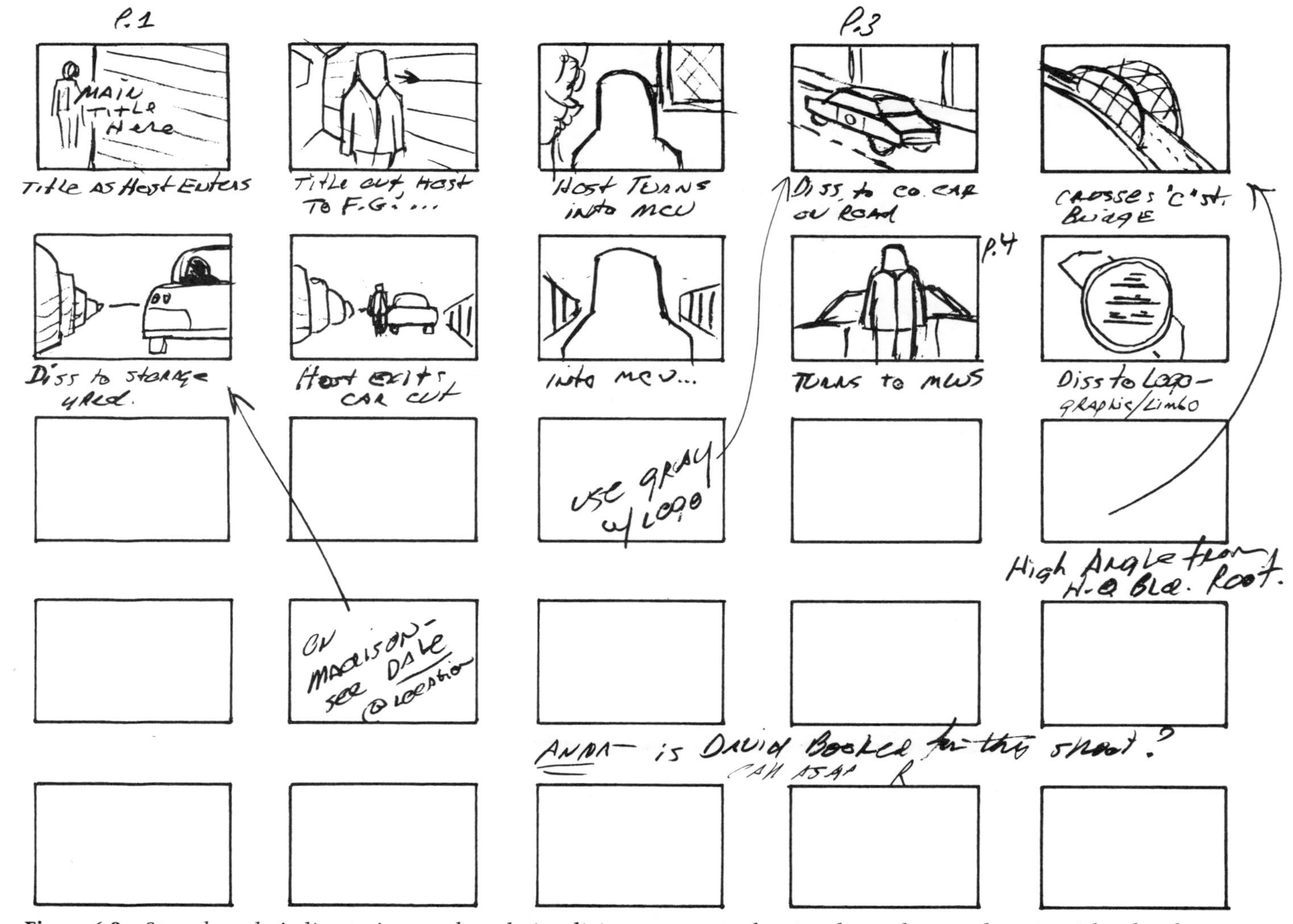

Figure 6.9 *Story board. A director's story board visualizing a segment that involves a host on location. The sketches are rough, but the director is not an artist. The line drawings simply suggest angles, focal lengths, and general framing. They allow both the director and the producer to analyze coverage to be shot.*

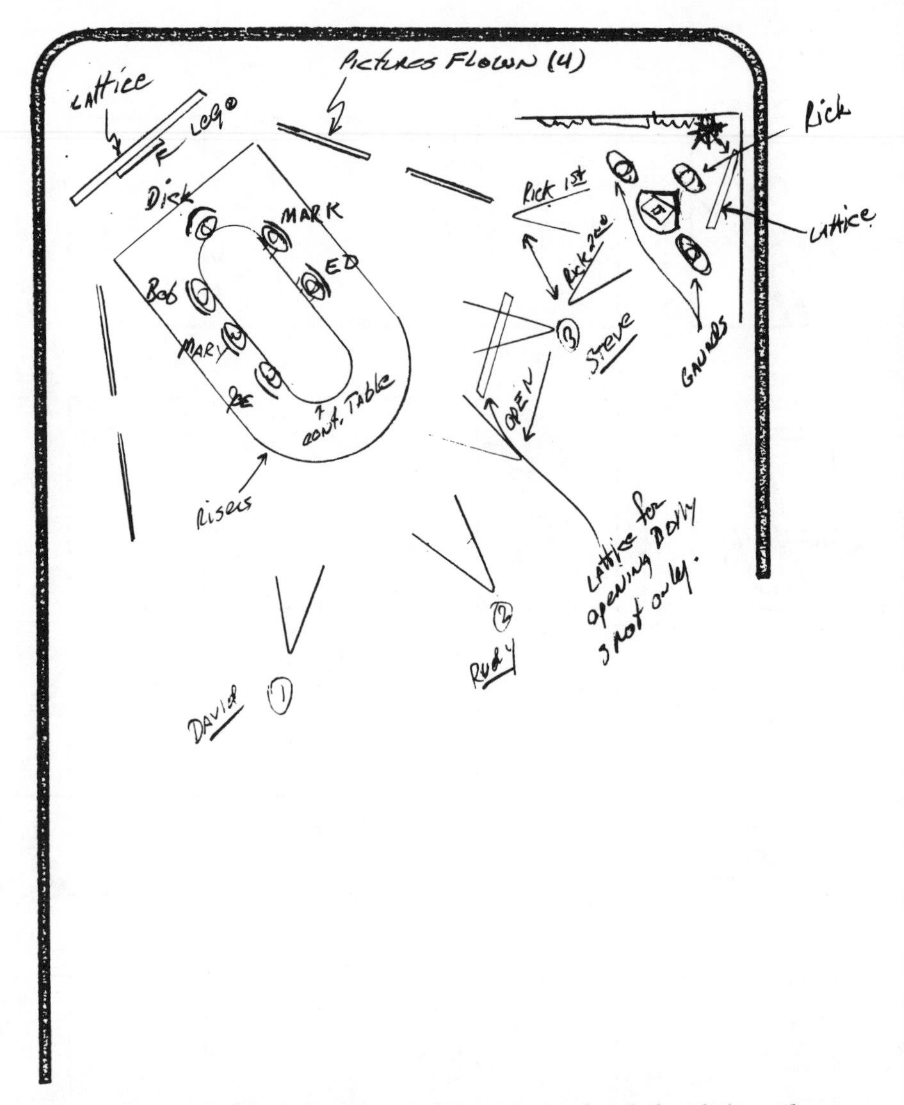

Figure 6.10 *Overhead diagram. A director's overhead sketch for a three-camera studio shoot. Two sets are shown. The first, at left, has six people seated around a table. Cameras 1 and 2 have a single basic position each. Camera 3 has an opening dolly shot. It also turns around at some point to shoot a person named Rick on the second set, at right.*

Second, blocking a dramatic scene is much more complex than blocking a simple host scene or an employee working. This type of blocking should be finalized on the set or location on the basis of how the scene plays out and its emotional content.

Blocking, however, is much like any other preproduction detail. If a good deal of thought and preparation is not put into it in advance, it usually leads to costly, frustrating delays in production.

REHEARSING

Talent rehearsals are rare in corporate television since they add up to extra time and actors fees. But when the project is a dramatic program, involving role-playing actors and complex camera blocking, it is a good idea to rehearse if the budget will bear it. If a full rehearsal on the set is impossible, a meeting between the actors and the director simply to read through and discuss the script can be very helpful. Rehearsing company executives is also very beneficial if they will allow you the time. Whatever type of rehearsal you use, the result will normally be better performances and a faster, smoother production.

In the case of multicamera studio shoots, technical rehearsals are common, usually on the day before production. In a *tech rehearsal,* as it is usually called, the entire crew is brought in and all elements of the shoot—camera angles, lighting, audio, props, and so on—are run through as a final check.

In any case, it is usually up to the director to pick the rehearsal day and up to the assistant director to do the scheduling.

CONDUCTING PREPRODUCTION MEETINGS

Preproduction meetings tend to follow the same rule of thumb as many other elements of production: The more complex the shoot, the more meetings are needed to stay on top of details as they develop. On most corporate productions, a single preproduction meeting the day before the shoot, or even on the same morning, is adequate.

Typically, the entire crew meets, scripts and schedules are passed out, and the director briefly discusses the shoot scene by scene. This type of meeting allows crew members to get an advance overview of what the director will want on particular shots and how he plans to get it. It also allows for crew input, which in many cases can make for a faster, more efficient shoot.

PREPARING EQUIPMENT FOR THE SHOOT

Following the preproduction meeting, or just before leaving for the first location, the crew checks over the equipment, repairs or replaces anything faulty, loads and gases up the trucks, gases the generator, and takes care of any other final details. This equipment checkout and repair process is an important one because of the nature of location television production.

During a typical location shoot, equipment is loaded and unloaded

Figure 6.11 *Engineer's home. The* bench, *as it's sometimes called, is where a video engineer spends much of his time. Cameras (such as the Betacam at right) and other pieces of equipment are constantly PM'd (given preventive maintenance) or fixed after being returned broken from location.*

at least several times. It gets carried, banged around, and piled up. It also endures a fair amount of bouncing around in the back of trucks. When this type of treatment takes its toll and a piece of equipment turns up faulty during production, there is often a tendency to put it aside and quickly grab another piece that happens to be working. The problem is that the piece that was set aside on the last shoot is sometimes left unrepaired and ends on *your* shoot next time. This is most common in the case of camera registration, lights with blown bulbs, and weak or dead batteries.

If your department has maintenance engineers who continually repair equipment and prep it before each shoot, you are less likely to end up with something faulty in the field. It's still possible, however, and finding the problem before you leave the facility can save you hours of frustrating down time on location.

FINALLY . . .

The equipment has been prepped and loaded, and all the other elements of preproduction have been attended to. Now, at last, you are ready (if the weather report doesn't predict rain) for production.

part three

Production

7 Production: Glamour or Grind?

OVERVIEW

Production is the process of recording all the sound and pictures called for in the script in a manner appropriate for use in the editing process.

Undoubtedly, there is an image of glamour associated with film and television production. Crowds gather quickly wherever a location shoot is in progress. Bystanders wonder aloud where the stars are or what commercial or movie is being shot. Children and adults alike attempt to slip into the background of a shot in hopes of later seeing themselves on television.

Anyone who has actually worked on a production, however, will attest to the fact that the word *grind* is a much more accurate description than *glamour.* Simply put, television production is *work:* long, hard, often frustrating work. At times, it also involves enormous amounts of pressure—especially for the producer and the director.

Before exploring how this work process actually takes place on a typical production, let's briefly review some basic but important preliminary information—the fundamental aspects of videotape recording.

THE VIDEOTAPE RECORDING PROCESS: FORMATS AND EQUIPMENT

The Process

Videotape recording is an *electronic* process, as opposed to film recording, which is a *chemical* process.

Film In film, an image is exposed to the emulsion on the film's surface for a specific, very short period of time. That exposure causes chemical changes in the emulsion, which, when the film is developed later, reveals a print of the image exactly as it existed for that instant. Frame after frame of these exposed images, projected rapidly on a screen, create motion pictures.

During production, the sound portion of a film scene is recorded separately on audio tape. This is accomplished on a special tape recorder, designed to run at exactly the same speed as the film camera so that the sound will later be "in sync" (i.e., synchronized) with the action.

Only in the final stages of the film postproduction process are these separate sound and picture elements combined to make the print we finally see projected on the screen.

Video In video, an image is first exposed to a number of electronic circuits in the video camera. These normally include three color tubes—one for the color red, one for the color blue, and one for the color green. During this process, the red, blue, and green portions of the picture are separated into individual electronic signals, which can be adjusted independently. These color signals and other aspects of the video image, such as *luminance* (light intensity) are processed by the camera and sent to a videotape recorder (VTR), where they are recorded at a rate of approximately 30 (actually 29.97) frames per second.

Unlike recording sound for film, recording the sound portion of a scene on video is *not* a separate process. Sound is recorded on an audio channel on the same videotape as the visual image.

Because electronic signals, unlike the chemical images recorded on film, do not require developing, videotape does not have to be sent to a laboratory before it can be viewed. A video signal can simply be fed from the camera into a television monitor with the proper attachment and viewed in real time as it is actually taking place. Once it is recorded on videotape, it can quickly be rewound and played back on the spot.

Obviously, the video method of recording sound and pictures is much more convenient and immediate than film. The editing process, as we will see in Part Four, is also much simpler in videotape. These factors have formed the basis for the phenomonal growth of video, both in the business world and in private households.

Now, let's look more closely at the different *formats* used in video and the types of equipment they require.

Formats

Most home video systems use the half-inch VHS format. *Half-inch* refers to the width of the videotape, and *VHS* (video home system) refers to the cassette type.

Three different formats are used in corporate and broadcast video: one-inch reel-to-reel, three-quarter-inch U-Matic, and half-inch camcorder.

Figure 7.1 *One-inch reel-to-reel VTR loaded with tape.*

One-Inch Reel-to-Reel One-inch reel-to-reel (Figure 7.1) is top-of-the-line, broadcast-quality video. It is what we see every evening on the network news and sitcoms. It provides very fine picture resolution; rich, true, color reproduction; and a very low signal-to-noise ("snow") ratio.

The one-inch reel-to-reel format is often used in corporate television studio productions. But because of its prohibitive price tag and the size and configuration of the equipment it uses, it is not the predominant video format, especially in productions shot on location.

Three-Quarter-Inch U-Matic Three-quarter-inch U-Matic (Figure 7.2) is a *cassette* format, rather than reel-to-reel. Its picture reproduction quality is not quite up to that of one-inch, but when combined with a good-quality camera, it is much better than the VHS format. In addition, because it is smaller and utilizes a cassette, it is much easier to use in the field than one-inch.

For many years, these qualities made the three-quarter-inch format the predominant standard for nonbroadcast video productions. The advent of half-inch camcorders, however, is quickly changing this.

Half-Inch Camcorders Camcorders use half-inch videotape cassettes in a VTR that is *built into the camera itself.*

The camcorder (Figure 7.3) has two important benefits. First, it eliminates the need for a separate VTR because one is built into the camera.

Figure 7.2 *Three-quarter-inch field VTR and power supply (on right). A foam-padded, hard-shelled carrying case is the typical means of transporting this type of equipment.*

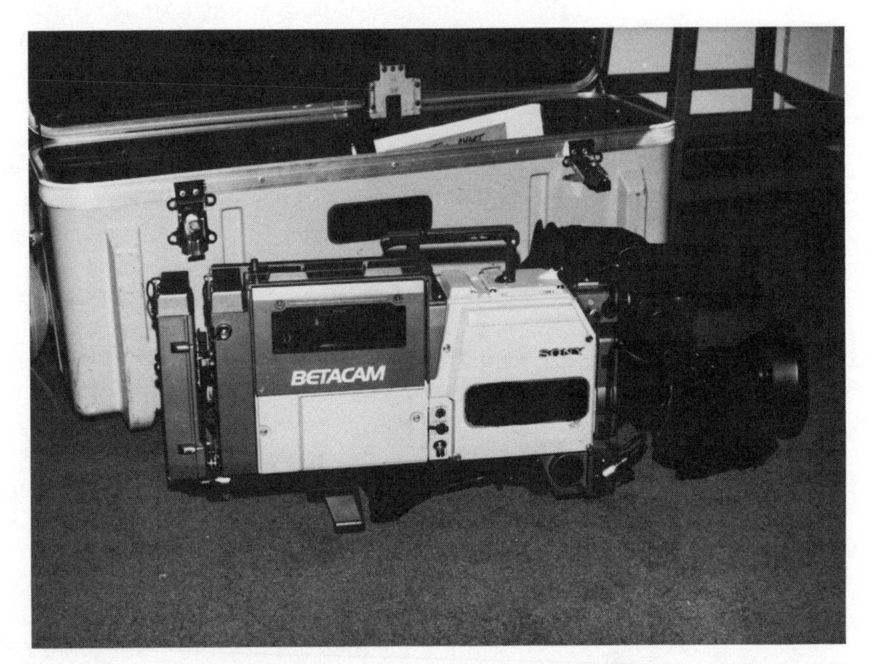

Figure 7.3 *Sony Betacam, a commonly used camcorder. The built-in VTR is the rectangular grey section on the back, upper part of the camera body. Vertical attachments at the rear of the camera are battery holders.*

This provides a cost savings as well as the ability to send out smaller and more mobile location crews. In addition, camcorders use a unique, high-speed, component recording system that approaches or even matches the picture quality of one-inch.

Equipment

In order to record in one or more of these formats, a video production requires certain pieces of electronic equipment.

Cameras A video camera receives a live image through its lens; converts it into a video signal; provides control over the signal primarily in terms of color registration, light intensity, and contrast; and finally sends the signal to a VTR. In the case of a camcorder, that VTR is attached to the camera itself.

As I've mentioned before, cameras used in corporate and broadcast productions normally come equipped with three tubes, one to process each primary color. These tubes are sometimes replaced by equivalent charge-coupled device (CCD) circuitry.

Single-tube cameras are sometimes used in corporate productions, but because the quality of the pictures they reproduce is normally considered inferior, they are the exception rather than the norm.

Videotape Recorders (VTRs) In production, VTRs receive *video* signals from a camera and *audio* signals from a microphone. They record each of these elements on separate parts of the same videotape. The video signal occupies the majority of the videotape surface and is recorded diagonally. The audio signal is recorded longitudinally on an audio channel near one edge of the tape.

Microphones Microphones send audio signals from the live source—often the actor—to the VTR. They come in three basic types: shotgun, lavalier, and handheld (Figure 7.4).

Shotgun microphones are long and slender (like the barrel of a shotgun). They are extremely *unidirectional* in their pick-up patterns. This means they pick up sounds best from one direction—straight in front of the microphone. Shotguns are normally used on *booms* (lightweight extendable poles) and are held *on axis* (in a direct line) with the actor's mouth, just out of the camera's range.

Lavaliers are usually very small condenser-type microphones, which clip to a tie or lapel. They are much less directional than shotguns and, therefore, record sounds well that come from most directions close to the microphone element. Lavaliers and other microphones that have comparatively broad pick-up patterns are often referred to as *omnidirectional*.

Pick-up patterns that are most responsive in two directions (usually opposite sides of the microphone head) are called *bidirectional*. *Handheld mics* are typically larger than lavaliers and usually also have om-

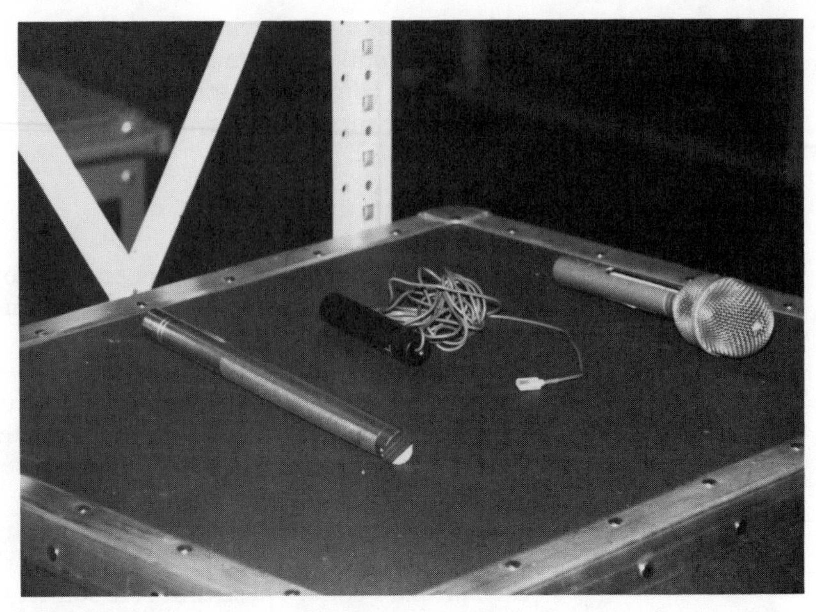

Figure 7.4 *Three types of microphones. From left to right: shotgun, lavalier, and handheld.*

nidirectional or bidirectional pick-up patterns. They are often used for on-camera host segments or are held by an interviewer in news-type pieces.

Although most audio signals are sent from one of these microphones to the VTR via audio cables, so-called *wireless* or *radio* microphone systems that transmit RF signals (audio signals converted to radio frequencies) are also common.

Wireless systems have a transmitter and a receiver. They can be adapted for use with each of the different types of microphones just mentioned and are normally used when audio cables would otherwise be seen in the picture.

Mixers Audio mixers are connected into the circuit between the microphones and the VTR. They provide two primary types of control over audio signals: volume control and the ability to combine signals. A typical field audio mixer may have four inputs and two outputs. Sounds from up to four different microphones could thus be fed into mixer input jacks, combined internally into one signal, and sent via one of the output jacks to the VTR for recording.

Monitors Two types of monitors are typically used in video production: standard video monitors and waveform monitors (Figure 7.6).

Video monitors are connected into the circuit to allow viewing of a scene either live or during playback after it has been recorded. This is usually done to analyze picture quality and light levels, as well as the actors' performances.

Figure 7.5 *A small field mixer. Three inputs can be seen at the left. Volume and other controls are on the front. Outputs (there are two on this model) are on the far side.*

Figure 7.6 *Waveform monitor (top) and video monitor (bottom) strapped to a "crash cart" for use on location. A three-quarter-inch VTR is on the next shelf down.*

Waveform monitors are usually connected into the circuit between the camera and the VTR. They provide a basic analysis of various elements of the video signal. The most important of these elements to the field production crew is normally luminance or light intensity.

On a waveform monitor, optimum light for recording is measured on a scale of 7.5 to 100 units. The standard color black, often referred to in video as the *pedestal* or *setup*, has a correct intensity of 7.5 units. White, at the opposite end of the luminance scale, has a correct intensity of 100 units.

If the white portions of a video picture are at 70 units on the waveform, the picture is said to be *underexposed*. If they are at 120 units, the picture, or at least that portion of it, is *overexposed*. Likewise, if the black portions are above or below 7.5 units, they are not standard blacks but, rather, *washed-out* or *crushed* versions of the color black.

Lights Although lights are not a part of the electronic system used in video recording, they are an important part of the overall process and thus deserve some basic discussion.

Production lights are so varied in size, shape, function, and output that it would take volumes to discuss each one in detail. Briefly, however, all production lights are used to do two things: first, provide enough light intensity to enable the system to record a video image

Figure 7.7 *Production lights. Barn doors, fixtures that cut off areas where no light is wanted, are attached to the Colortran 1K at the left. On the right is a Mole Richardson 1K without barn doors. Both are Fresnel-type lights with glass lenses over bulbs.*

Figure 7.8 *Lighting console. A console such as this is used to patch studio lights to dimmers. For example, a studio light plugged into outlet 55 on the grid above the set can be assigned control at dimmer 2 on the console. The light can then be quickly raised or lowered in intensity as required.*

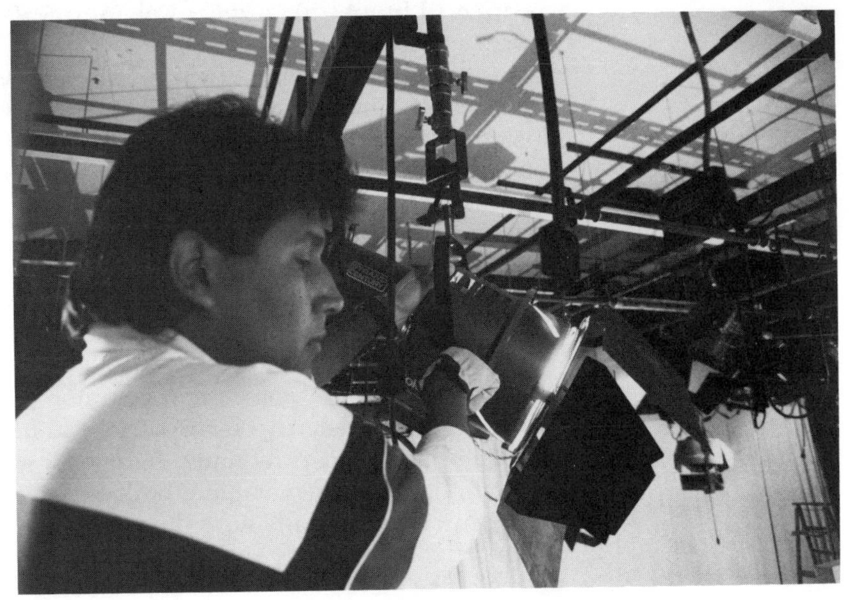

Figure 7.9 *Gaffer setting a 1K (1,000-watt) light. Note the barn doors on the front of the light. The ring around the middle section of the light is used to* spot *(focus) the beam or* flood *(defocus) it.*

within the acceptable standards as measured on the waveform monitor, and second, improve the aesthetic quality of the picture through the use of highlights and patterns, modeling with shadows, and maintaining certain color hues or temperatures.

Some lights have adjustable beams, which can be spotted or flooded for different intensities and areas of coverage. Others can provide direct, hard light or a softer bounce or diffused light. Still others provide different color temperatures.

Color temperature refers to the color of light on a standard scale. This scale measures the color radiated by a standard *black body* (such as a carbon filament) when it is heated to different temperatures on the Kelvin scale.

As an example, when exterior daylight is recorded on videotape, it is very blueish in color, with a color temperature of approximately 5,500 degrees Kelvin. Arc-type and halogen-metal-iodide (HMI) lights provide this same blueish light and are thus used to imitate or mix with daylight. Interior light, on the other hand, is much more orange, with a color temperature of approximately 3,200 degrees Kelvin. This temperature is usually provided by tungsten/halogen light sources (a tungsten filament sealed in halogen gas inside the bulb) and is used for interior shooting.

Unless it is done with skill and purpose, mixing lights of different color temperatures can result in pictures that look obviously lit and unnatural.

Gels, colored films placed over the light source, are used to change the color temperature of the light being emitted. A tungsten/halogen light, for instance, with a specific grade of blue gel placed over it, will be corrected from 3,200 degrees Kelvin to 5,500 degrees, thus making it usable as an exterior light source.

Gels are also used simply to color a light source for visual effect. One example would be an orange gel placed over a light to create the illusion of light from a fireplace out of the camera's frame. Cloth strips dangled and moved in front of the light can be used to create the flickering effect.

Other light accessories include *scrims* and *nets,* which, when placed over the light source, decrease its intensity; *flags,* which cut off the light from certain parts of the picture; and *silk, spun,* and *frost* materials, which diffuse or soften the light falling on a scene or object.

Other Equipment Some other common pieces of equipment used in production include tripods or *sticks* and *heads,* as they are called, on which the camera is mounted. Tripods, usually made of hard wood or tough, lightweight metal, are built to provide sturdy, rigid camera placement. Heads usually have fluid workings that allow very smooth pan and tilt movements with little or no camera shaking. *Spreaders* are spokelike pieces, usually metal, that are attached to the feet of the tripod to keep its legs locked in a spread position.

Dollies, on which the camera can be rolled forward or backward (called a *dolly in or out*) or sideways (called a *truck left or right*) come

Figure 7.10 *Flag in place. A flag, in the upper right-hand corner, is set in place on a C stand. The flag will block off some portion of a light source above and behind it.*

in a variety of forms. Doorway or western dollies are essentially simple wooden platforms with four wheels and a handle (Figure 7.12). The camera, mounted on its tripod, is secured onto the platform. The camera operator and perhaps a focus puller also mount the dolly. As the shot is being performed, the dolly is pushed or pulled, normally in one direction, by a grip. With special wheels and sections of tubular track on which the whole system can ride, the dolly move can be made very smoothly.

The more expensive dollies, Elemacs, Fishers, and Chapmans, are made of heavy metal. Most of these dollies also have *pedestal* or *ped* controls for up or down movements of the camera and *crab* mechanisms that allow the dolly to move in arcs, trucks, dollies, and combinations of these all in one shot.

Video and audio cables (Figure 7.13) are also standard items on any shoot that is recording sound as well as pictures. They are very different in makeup and, of course, serve different purposes. One cannot be substituted for another.

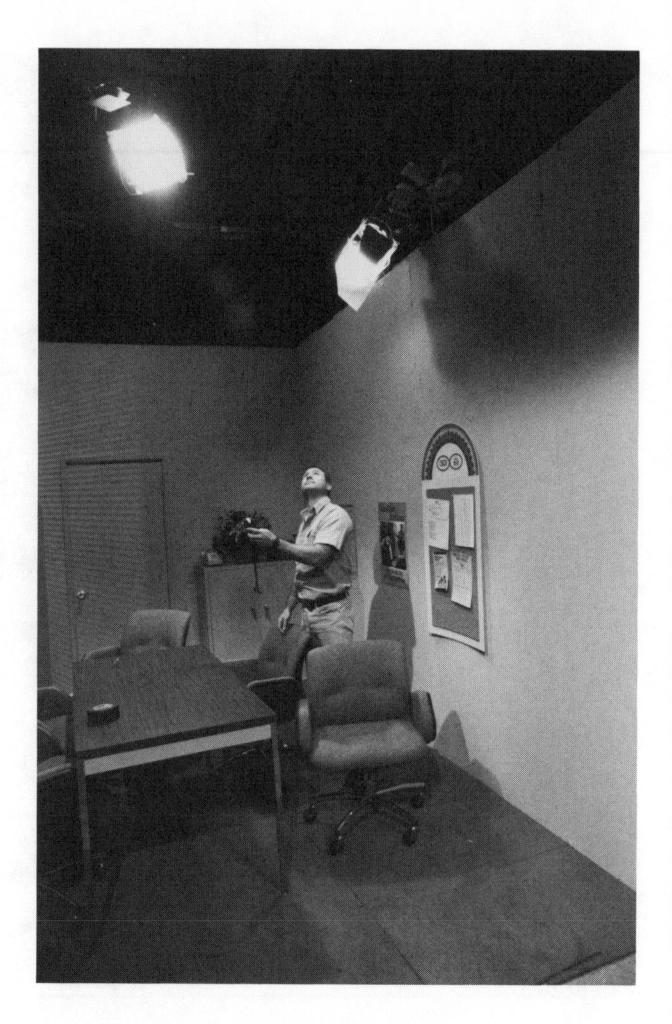

Figure 7.11 *Tweeking. The light above is adjusted according to the reading on a light meter held by the person below. Note that the front of the light has been fitted with* spun, *a gauzelike material used to soften the beam.*

Batteries, generators, A/C cables, and power supplies provide power where none exists. Normally, the only pieces of production equipment that cannot be battery-operated are lights. These require a generator because they operate at high voltage and amperage levels. In some cases, small lights are operated on battery belts or battery sources, but they usually can be used only for very limited periods of time.

C stands are three-legged metal stands used most often to hold flags or nets in front of lights. *Apple boxes* are wooden cubes of varying thicknesses (one-half apple, one-quarter apple, etc.) used for a variety of propping up or elevating jobs.

Finally, *gaffer's tape* is used for everything from dulling hot spots to sticking mics in collars. It looks just like duct tape, but with a rough, dull gray finish. Gaffer's tape is considered an essential *expendable* (an item that can be used up) on most shoots. Legend has it that gaffer's tape will stick just about anything to anything else.

Now, let's look more closely at the production process itself. There

Figure 7.12 *Doorway dolly. Special track wheels lie on the platform where the camera* sticks *(tripod) are normally placed. Chrome sections of* track *can be seen lying behind the dolly.*

Figure 7.13 *Production cables. From left to right: Standard video with BNC connector, standard audio with XLR connector, and a typical A/C (alternating current) electrical cable.*

are two types of television production: location and studio. Both have their benefits and their drawbacks.

LOCATION SHOOTING

As a rule, a film or videotape crew goes on location only when they can't achieve what they must in the studio. The reason is simple—shooting on location is far more time-consuming, much harder work, and far riskier because the crew is at the mercy of so many unpredictable factors.

On the other hand, only location shooting can actually transport the audience to, say, an actual job site. No matter how well lit a studio set may be, on videotape it tends to look like a set instead of the real thing. Footage shot on location has a truer feel to an audience, an important factor in the credibility of the final program. For these reasons, not to mention the simple fact that a company may not have a studio or be willing to rent one, shooting on location is a common occurrence for many corporate video units.

Single-Camera Style

In some cases, a studio setup, including several cameras, a switcher, and so on, is brought on location. Most location shooting, however, is done in what's called *single-camera style,* because it makes use of a single camera, which is moved from position to position for each angle—the same way films are shot. In this type of shooting, each actor's lines and action must be repeated for each angle, and this *coverage* of the scene is cut together later in the editing process.

Typically, a director on location will first shoot a wide or master shot of the action. Next, he will shoot the same action in, perhaps, medium shots, taking in much less of the overall scene. He may then choose to shoot close-ups of certain parts of the same action.

Later, when the editor gets the footage, she will most likely start by using several seconds of the master shot. At that point, she may cut to the medium shots and then, several seconds later, to the close-ups. Aesthetically, this allows viewers to establish quickly where the scene is taking place and then move in closer for emphasis on the actors and key pieces of action.

Key Elements

From a producer's viewpoint, there are several key elements of location shooting. Locations provide a more credible look than studio footage, but location shooting also involves much more work. In addition, footage shot on location in the single-camera style is assembled after the fact from repetitions of the same action shot from different angles and focal lengths. This, as we will see, means more editing time but also provides more flexibility than studio shooting.

STUDIO SHOOTING

Studio shooting takes place in a much more controlled situation, and therefore a much more consistently productive one. In a studio, a sudden gust of wind cannot ruin a host's monologue; cars do not come by and honk; the lights are consistent and controllable, unlike the sun and clouds, which are constantly changing in ways that affect the continuity of scenes being shot. There are no breakers to trip, no generators to run out of gas suddenly, no bystanders to step into a shot and wave, and no equipment to turn up missing when you're thirty miles from the studio.

It is for these reasons—ease and control—that studio shooting is usually preferred over location shooting.

Multicamera Style

Although studio shooting can also be done with a single camera, in television it is often done in *multicamera style.* In this style of shooting, the director, rather than moving a single camera, simply switches the video images going onto videotapes or on the air by choosing the signals from any of *several* cameras out on the studio floor.

For example, he may fade up from black on camera 1, which is on a wide shot of the set and actors. Once their conversation starts, he may want a medium shot of, say, the leading man. The operator on camera 2 has a medium shot, so the director switches to or *takes* camera 2.

Obviously, in this type of shooting the program is basically edited or covered on the run. For this reason, studio shooting cuts down substantially on the editor's time in postproduction.

Disadvantages of Studio Shooting

Studio shooting, however, does have four primary negative characteristics. First, as previously mentioned, a studio set never looks quite as credible as the actual location. Second, studio shooting requires the time, personnel, and material resources to have sets designed and built in the days prior to shooting. Third, the ability to manipulate multi-camera footage in editing can be minimal. Finally, if your company does not have one, well-equipped studios are very expensive to rent—usually around $300 to $500 per day.

Now, with this background in mind, let's go on location and into the studio.

8 A Day on Location and a Day in the Studio

LOCATION PRODUCTION

Location shooting days usually begin between 5:00 and 7:00 A.M. The crew arrives at a predetermined call time and either loads the trucks, or, if this has been done the evening before, does a final equipment check and heads for the first location. Once the crew has arrived at the location, which we will assume is a large office area, the initial setup begins.

Setup

During setup, several things happen at once. The grips and gaffers begin to unload while the assistant director (AD) gets the actors situated and makes the clients comfortable and attends to any other details that arise. The director and his director of photography (DP) look over the area and discuss the first shot.

The director explains where he would like the camera positioned, what the action will be, and how close or wide he would like the shot framed. The two also discuss how the scene should be lit—with a soft or hard look, contrasty or flat (lacking shadows), and so on.

Once the DP has a clear idea of what the shot will involve in terms of lighting and blocking, he calls on his gaffer and explains what lights he would like to use and where they should be positioned.

Typically, he will want to light a scene with a *key* light somewhere close to the camera that will appear to be the main source of light. This might be a *2K Mole* (2,000-watt Mole Richardson) Fresnel—a light with a glass lens that is capable of spotting or flooding. It might be placed roughly 5 feet to the right of the camera, about 20 feet back from the talent, at a height of about 7 feet.

The DP will probably also want a somewhat dimmer and softer source of light on the other side of the camera as a *fill*—to fill in the harsh shadows created by the key light. For this job, he might select

Figure 8.1 *Inside location. A company employee is videotaped in a company conference room. The large soft light* (softy) *at the upper left provides the key light. The circular reflector held by the gaffer uses the same light to bounce fill light onto the opposite side of the subject's face. The sound engineer is ready with a shotgun mic and a portable mixer on the table beside him. The camera is out of the frame, behind the clipboard.*

a *2K softy*—a light that is less focused and is bounced out of a reflective hood toward the talent.

If both these lights are placed roughly the same distance from the actor being lit, they will produce a mild contrast ratio (light/shadow ratio), which is typical of corporate programs.

Finally, the DP will ask for a back light on the actor, which will act to make her stand out from the background. A *1K Mole* placed behind and above her head would do well for this. Depending on how wide the shot is and what else is seen, the DP may want more lights for the background or other objects.

The gaffer (Figure 8.2), whose job it is to know the electrical values and capacities of lights, fuses, breakers, and so on, now talks to the grip about where the lights will be plugged in, how many can go on a particular circuit, and so on. The two then begin laying out A/C cables and setting lights (Figures 8.2 and 8.3).

Meanwhile, the engineer-in-charge (EIC) is setting up the VTR, a video monitor, and a waveform monitor. Since in this case she is the sound engineer also, she sets up the mixer and the microphone to be used, too.

Since the teleprompter and the camera will end up mounted on the tripod as a single unit, the teleprompter operator has now begun working with the DP to set this up.

During this time, the director has discussed with the AD what shot

Figure 8.2 *Gaffer adjusts reflector. Lights are not always readily available. Here a gaffer sets a reflector. The smooth reflective surface bounces light (usually sunlight) onto the subject. The opposite side of the reflector is dulled for a softer look.*

this will be and which scene it belongs to. During this conversation, they will no doubt both refer to each other's scripts. They will discuss whether it is a master scene that will be shot first, which actors or employees will be involved, and probably the type and extent of coverage the director plans to shoot. On the basis of this conversation, the AD prepares a slate with the proper scene number on it, finds a good spot from which to view the action, and handles any other details that come up.

The director now turns his attention to the actress, in this case the hostess. He explains what action she will perform and discusses the type of *read* he wants—whether serious, light, formal, or some other tone. Once the actress is comfortable with what the director wants, she begins to rehearse.

Finally, the lighting has been *tweeked* (adjusted) and looks good. Prior to this, the DP and the EIC have probably met at the waveform monitor several times to discuss and resolve video-level problems such

Figure 8.3 *Location sound. A shotgun microphone inside an elongated zeppelin is mounted on a boom and pointed toward the subject. A reflector can be seen at upper left. The camera battery pack beneath the tripod provides power. The talent, a company employee, is behind the reflector. The director, to the right of the camera operator, watches the action.*

as *hot spots* (overexposed portions of the picture) or other lighting problems. No doubt the DP has placed various flags and/or nets on C stands to reduce light in certain areas in order to solve some of these problems.

The actress now appears to be reading well. The lavalier mic has been taped just under her collar, and the EIC has had her read several lines to verify that her level is producing a good audio reading on volume units (VU) meters at both the mixer and the VTR.

The camera and prompter have been mounted together on the tripod, and the camera is white-balanced and ready to go. *White balancing* properly references the camera's circuitry to the color white, thus making it able to reproduce all colors accurately. White balancing is accomplished by pointing the camera at something white, often the back of a script page, and activating the white balance circuit.

The DP now switches the camera to the *bars* circuit, which sends a standard pattern of colored bars to the VTR. At the same time, the EIC hits a switch on the mixer that generates a 1,000-cycle tone. With both the tone and the bars being sent to the VTR, the EIC puts the machine into record and *lays down* approximately one minute of this.

This *bars and tone,* as it is usually called, will be laid on the *head* of every tape that is recorded. Later, it will provide a reference to set up the duplication equipment used to duplicate the footage. In this way, consistent levels of light intensity, color value, and sound can be assured between segments of recorded material.

Roughly an hour has passed since the crew arrived. Finally, the director calls for a rehearsal on camera.

Rehearsal

The AD now calls for quiet, and the director says, "Action." Generally, this first rehearsal is rough, in several ways. If the DP has a *move,* for example a dolly or zoom that must start and end on certain lines, he may need a few run-throughs to get the timing just right. In addition, the actress's performance may have rough edges, or the background action may not be just right. After several rehearsals, then, when all the bugs in the shot are taken care of, it's time for take one.

Take One

Again, there is a call for quiet. This time, however, the director tells the EIC to "roll tape." The EIC does so and, in a moment, tells the director she has "speed." This means the tape machine is up to speed and is now recording.

The grip now holds the slate he has received from the assistant director in front of the camera. On it might be written: Scene 15, Take 1, Roll 1, the title of the production, the date, and the names of the director of photography and the director. (We will see the importance of the slate in the section on postproduction.) The EIC tells the grip to "mark it," and the grip says, "Scene fifteen, take one." The scene now has a visual and audio *signature.* After the slate is pulled away, the director pauses for a moment and says, "Action." The scene plays out.

In all probability, something was not right. Perhaps the microphone rustled when the actress walked, or the zoom was late, or someone in the background looked at the camera, or a line was blown. It could be any number of irritating minor occurrences; in any case, it means that the process must be repeated for a take two. After anywhere from two to perhaps fifteen takes, everything plays out to the director's satisfaction, and he says, "That's a buy."

During all this, the AD has been noting the number of each take and writing a brief description of it—including anything that went wrong—on her master script. She will continue to do this for every shot of the production. Now, however, when she hears the director

Figure 8.4 *Crew on location. The director, in the foreground, watches the monitor (just under his clipboard). The camera operator rolls his Betacam, a self-contained camcorder. The grip stands by as the talent (a company employee) exits the truck.*

declare that scene 15 is a *buy,* she draws a circle around the number of that take.

Figures 8.5 and 8.6 are examples of master script pages. As with the slate, we will see how the master script becomes important in the section on postproduction.

Once this first scene is recorded on tape, the director may or may not want a close-up or medium shot of the same action. If he feels it is necessary, he will ask the DP to reframe the same scene with a new focal length or angle and explain to the AD what he is about to do. A new slate will then be produced.

If the new shot is an *alternate* enactment of the scene, the slate will be marked A15. This indicates to the editor a different version of a scene. For example, the hostess might gesture to the left instead of to the right. If, on the other hand, it is a close-up of the *same* action or a part of that action, it will be marked 15A. This indicates an *insert* into the existing scene rather than an alternate version. An example of this would be a close-up of the hostess's hand picking up a tool just as she did in the master scene. A second insert of a different part of the scene would be 15B, a third 15C, and so on.

Whatever the slate marking, the same recording process will take place again and very likely several more times before the director buys it.

The Strike

After perhaps an hour and a half from the time the crew first arrived, the director will have what he wants and will instruct the crew to *strike* the location. That means it's time to pack everything back up and move to another location to do the same thing over again. If the next location is nearby, it may mean simply moving some equipment a short distance and relighting. If it is not close by, it will probably mean putting all the equipment back into cases and boxes, loading the trucks, and driving to the new location.

This process will repeat itself perhaps five, ten, or even fifteen times during a typical ten-hour shooting day, again depending on the complexity of the shots, their distance from each other, and any other number of complications that tend to arise during location production.

Roughly ten hours after they began, the crew, thoroughly exhausted, will roll back into the yard. If this was only a one-day shoot, all the equipment may then need to be unloaded before anyone leaves. If it is a two-day shoot or longer, the equipment may stay loaded in the trucks, and the crew will head for home immediately.

The AD will probably stay behind to catch up on paperwork or notes left unattended to in the heat of production. The director will also stay behind and probably look at *dailies*—shots recorded that day. The DP and the producer will probably also attend this dailies review session to be sure the footage brought in is acceptable. Later, the director will take his script book home to look over what he has accomplished so far and the next day's schedule.

As is obvious, location television production means hours of loading, unloading, rehearsing, lighting, tweeking, and so on to produce minutes or even seconds of usable footage. The experienced corporate producer, recognizing this fact, learns to weigh the positives and negatives of different types of shooting against the needs of the project and the resources at hand. Some common considerations are: Does it need to be shot on location? If so, can it be a location close to the facility or offices? Can what the script describes as multiple locations be combined into one location, which is re-dressed to look like others? What are the costs for the locations? And so on.

The experienced producer also knows that considerations of the number and types of locations that arise in preproduction or production are probably weeks too late to be economically helpful. Ideally, these decisions should be made in the script stage, when the program is being designed and written.

STUDIO PRODUCTION

Studio shooting tends to be much less strenuous than location shooting once the sets have been built and lit. It also tends to be much more

3 R1 6/13	MASTER: WIDE SHOT. HOST GETS UP FROM DESK, WALKS TO FRONT. Z.I. ON GRAPHIC.
1.	N.G. HOST MISSED LINE
2.	N.G. ZOOM SHAKEY
3.	HOST STUMBLED
(4)	GD -
A3 R-2 6/13	ALT. MASTER. NO ZOOM IN TO GRAPHIC
(1)	GD -
3A R-2 6/13	SAME ACTION AS MASTER IN CLOSE UP
1.	N.G. HOST MISSED LINE
2.	N.G. "ROUGH" READ
3.	STILL ROUGH
(4)	GD. -
4 R-4 6/14	MASTER: MED. SHOT, JENNIFER @ DESK ON PHONE W/ CUSTOMER
1.	N.G. POOR PERFORMANCE
2.	N.G. MIC RUSTLE
3.	EMPLOYEE STOOD UP IN B.G. (1ST. PART GD)
(4)	GD. -
4 R-4 6/14	C.U. SAME ACTION
(1)	GD-
4B R-4 6/14	HOST V.O.
(1)	GD-

Figure 8.5

A master script page (Figure 8.6) with the AD's notes for the coverage on the back of the preceding script page (Figure 8.5). Vertical lines and scene numbers indicate how much and what type of coverage has been shot for those parts of the script. The squiggly line indicates audio recording only. The opposite page (notes) indicates scene numbers, reel, date, and a description of the

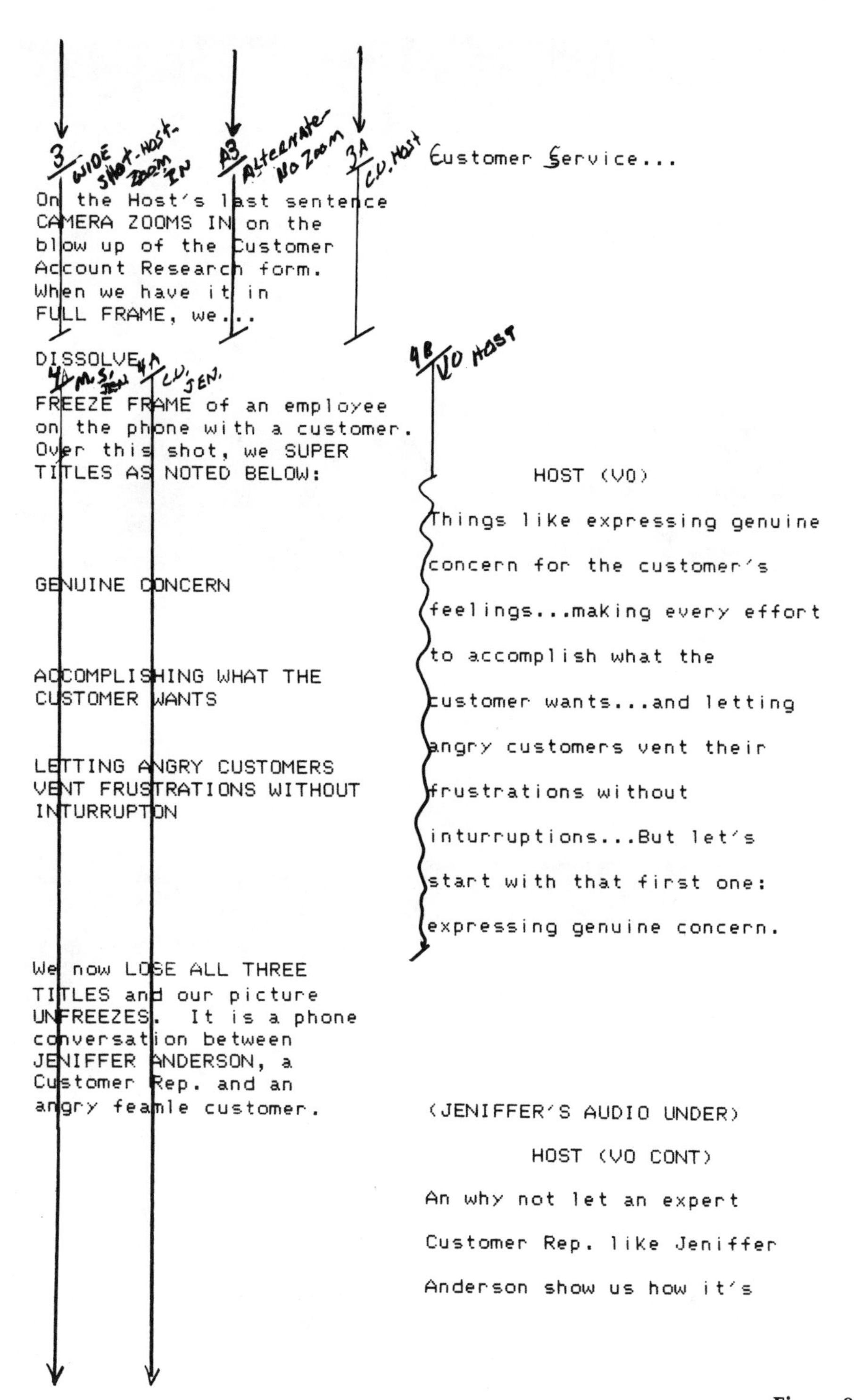

Customer Service...

On the Host's last sentence
CAMERA ZOOMS IN on the
blow up of the Customer
Account Research form.
When we have it in
FULL FRAME, we...

DISSOLVE

FREEZE FRAME of an employee
on the phone with a customer.
Over this shot, we SUPER
TITLES AS NOTED BELOW:

HOST (VO)

Things like expressing genuine
concern for the customer's
feelings...making every effort
to accomplish what the
customer wants...and letting
angry customers vent their
frustrations without
inturruptions...But let's
start with that first one:
expressing genuine concern.

GENUINE CONCERN

ACCOMPLISHING WHAT THE
CUSTOMER WANTS

LETTING ANGRY CUSTOMERS
VENT FRUSTRATIONS WITHOUT
INTURRUPTON

We now LOSE ALL THREE
TITLES and our picture
UNFREEZES. It is a phone
conversation between
JENIFFER ANDERSON, a
Customer Rep. and an
angry feamle customer.

(JENIFFER'S AUDIO UNDER)

HOST (VO CONT)

An why not let an expert
Customer Rep. like Jeniffer
Anderson show us how it's

Figure 8.6

scene. Beneath this are listed the various takes and what went wrong in each one. The buys are circled. This system of using the back of the preceding page to make notes works well in the field when things are moving fast. Sheets are added if notes get too extensive.

Figure 8.7. *AD and director. The assistant director (AD) goes over last minute details with the director.*

Figure 8.8 *Powwow on the set. Prior to the technical rehearsal, the producer, director, client, and AD gather to check the set, dressings, props, and so on.*

Figure 8.9 *Getting set. An operator checks out his camera prior to the shoot.*

controllable and hence more efficient. This, of course, is the reason sound stages were originally built.

On the morning of the studio shoot, the crew will have a call time roughly one hour before the director plans on rolling tape. This allows time for the cameras to be turned on, registered, white-balanced, and so on. It also gives the technical director (TD) time to put a reel of videotape up on the production recorder, turn on and set up all the house equipment, recheck video and audio levels, and help with camera adjustments.

During this time, the actors will have arrived and probably will have been shown to the dressing room to get made up and dressed. The director will most likely visit with them, pump them up a bit for the day's performances, and discuss and choose wardrobe for the various scenes.

The AD, meanwhile, will be preparing her script, helping the actors with their needs, and working with the crew and clients or other dignitaries who are also arriving. And the camera people, besides preparing their cameras, are probably tweeking lights at the insistence of the DP.

Rehearsal

When everything is set, the actors will be called to the stage to discuss the first scene with the director. The process is basically the same as on location, with the exception mentioned earlier: The director will

Figure 8.10 *Warming up the talent. The director warms up a company employee to the studio environment.*

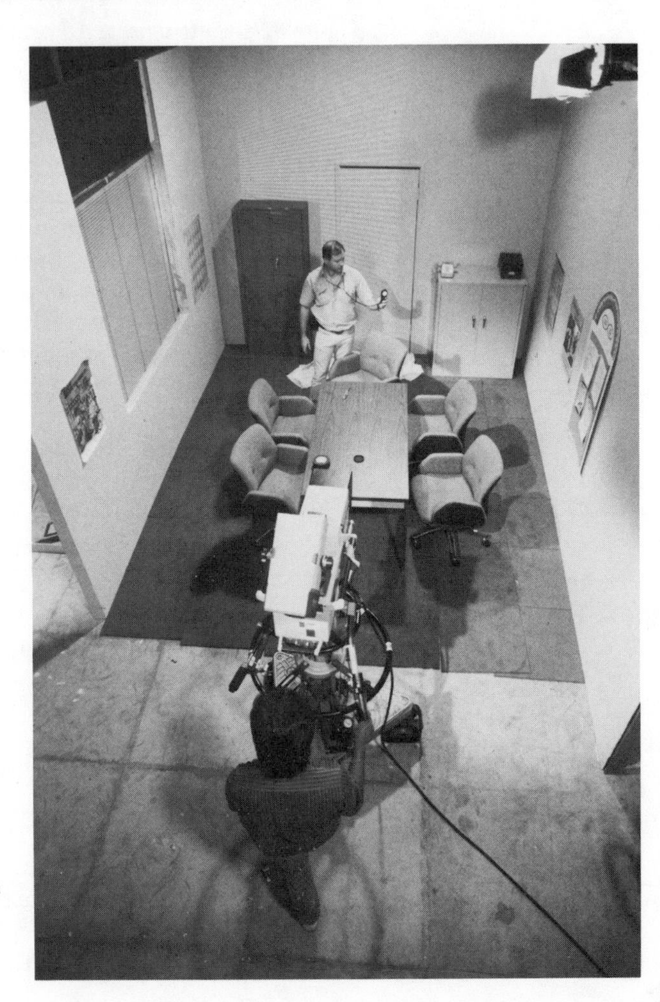

Figure 8.11 *Tweeking. Once the lights are basically in place, a camera is brought in to start showing what the exposure actually looks like on the monitor.*

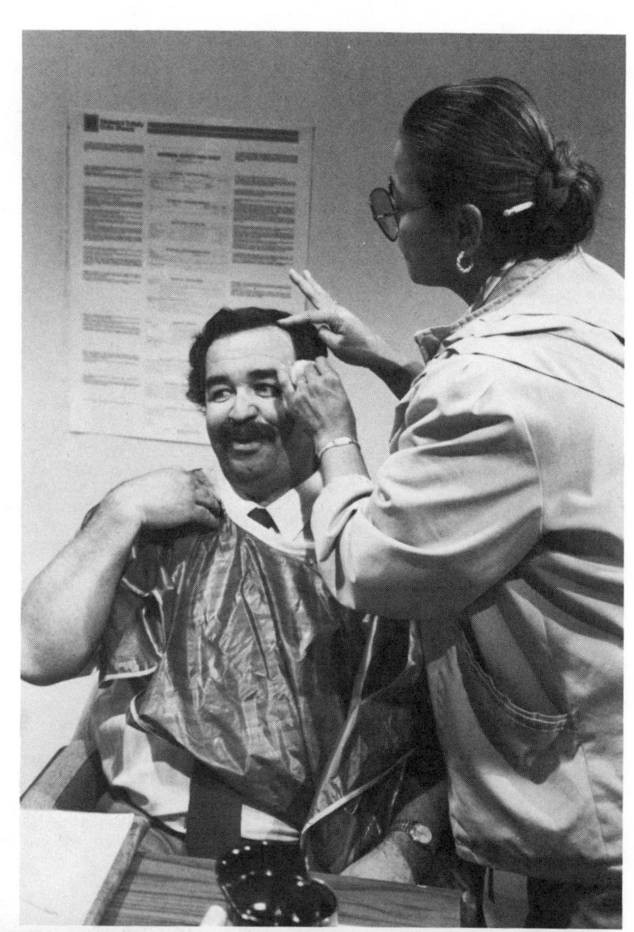

Figure 8.12 *Getting dusted. A company employee gets "dusted" to keep the shine off. To accomplish this, a light coating of powder is applied to his forehead by the AD.*

Figure 8.13 *Framing up. During the tech rehearsal, camera operator frames up a shot for the director, who is standing just behind him. It is during this time that all the bugs are worked out of the shots. When the tech rehearsal is over, all crew members should be totally prepared for the following morning's production.*

Figure 8.14 *Final adjustments. Just prior to rolling tape, final adjustments are made to a microphone concealed beneath a collar. Rustle from hidden mics is a common headache. Meanwhile, the client looks over the set to be sure everything is as it should be.*

Figure 8.15 *Master control (before the "heat"). Just before production, the control room, or* booth, *as it's called, looks orderly and neat—a condition that will surely change when the heat is on. Seated from left to right will be the teleprompter operator (the prompter camera is the small back unit above the chair pointed down at the white conveyor unit), the AD, the director, the technical director (TD), and the video control (VC). The VC adjusts camera levels as production is in progress. The lower bank of monitors display different video sources. Three are from cameras out on the studio floor. The preset monitor is the left-hand one of the two large monitors. Program—what's actually going on tape (or on the air)—is the large one on the right.*

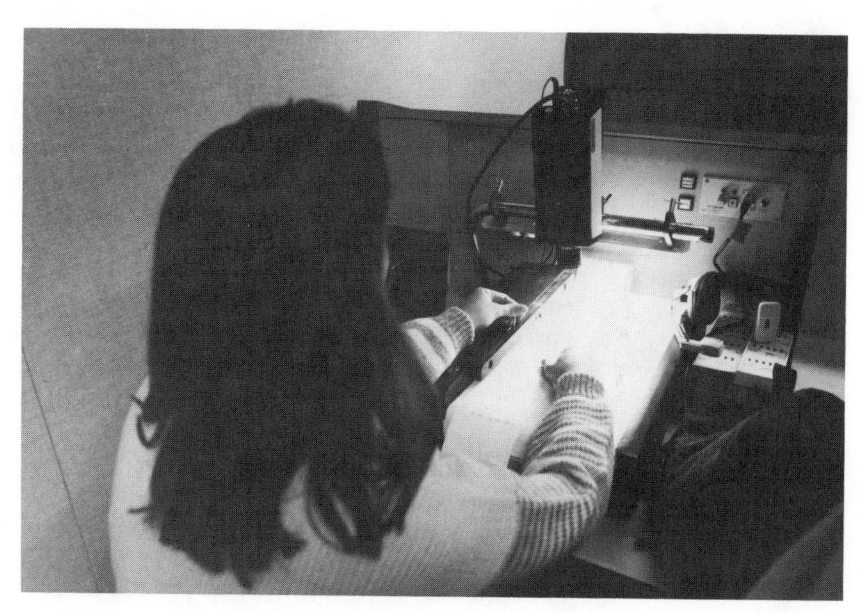

Figure 8.16 *Teleprompter operator during studio shoot. A small video camera (black and white rectangular unit) looks down on the host's lines, which have been typed on a long, slender roll of paper. The image is then sent to a monitor that is actually mounted on the front of a studio camera. The image from the monitor is reflected off a one-way mirror placed over the camera's lens. In this way, the hostess is able to read her lines by looking directly into the camera lens rather than off-camera at cue cards.*

Figure 8.17 *Betacam ISO. A camera operator using a Betacam (self-contained camcorder) as an ISO (isolated camera) in a studio production. ISO means that the camera is recording to its own individual tape recorder. ISOs are often used to double-cover segments being switched live. In this capacity, it provides backup insurance. If the director switches to camera 3 and, in doing so, misses some crucial piece of action, it will still be recorded on the ISO camera and can later be edited into the program.*

now have a number of cameras at her disposal, so she will be able to shoot all her coverage of the scene with a single pass—a single good pass, that is.

Once the scene has been walked through a few times and the director is satisfied that the action and performances are close, she will generally leave the stage and go to the *booth*—an adjacent control room equipped with banks of monitors, scopes, controllers, and a video switcher. Here she will take a seat between her AD and her TD and, just as on location, she will call for rehearsals and, finally, "Action."

Switching Live-on-Tape

This time, as the actors go through their paces, the director watches the camera angles and focal lengths she has on the monitors before her and calls out her takes to the TD, who switches the show "live-on-tape."

During this process, everyone on the crew is connected via a private line (PL) on headsets. All crew members act on the basis of the director's commands.

The following is a list of typical director's calls during an opening portion of a show. Also shown is the action that takes place and who does what:

Director says . . .	*. . . and this happens*
Roll tape.	The TD puts the production recorder into "record." He acknowledges the machine has "speed."
Slate it.	The TD switches to a camera or electronic source with the proper slate. He or the audio engineer call out the scene and take number into an open microphone.
Ready to fade up on one.	The TD sets camera 1 to preset on his switcher and grasps the *fader bar.*
Fade up.	The TD pulls the fader bar, and the picture on camera 1 appears on the Program monitor. This is the image being recorded on tape.
Lights up, please.	The floor manager slides one or several *dimmers,* which cause the lights on the stage to "come up."

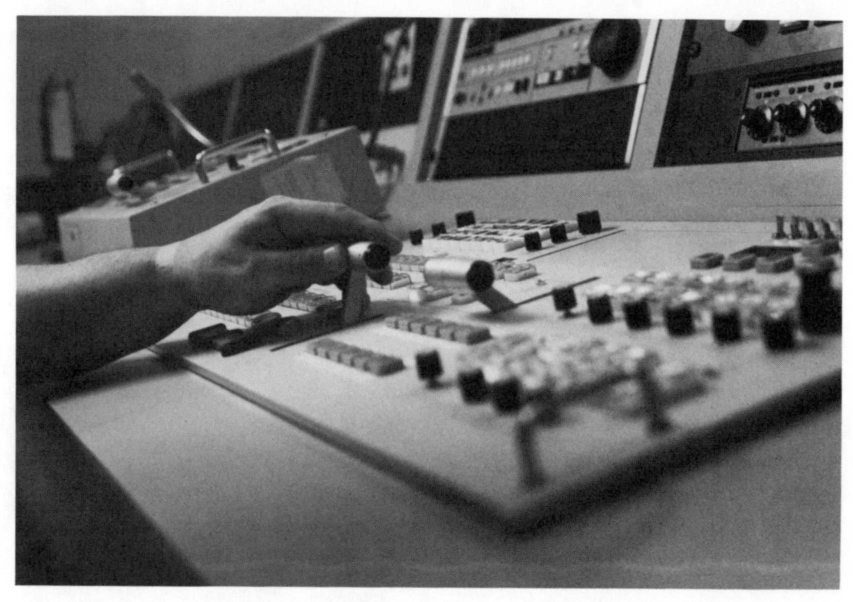

Figure 8.18 *Fade in. The TD pulls the fader bar on a Grass Valley switcher. A digital effects generator (DVE) can be seen in the background.*

Mic up and . . .	The TD or sound engineer "opens" the proper mic for the actor about to speak. Sound is now being recorded.
Cue talent.	The floor manager points a finger at the actor whose job it is to walk on and start the scene. The actor does so.
Camera 2, give me a close-up, please.	The operator on camera 2 zooms in quickly to a close-up of the actor she is on.
Ready two.	The TD sets camera 2 to the preset position on his switcher.
Take two.	The TD "switches" to camera 2, and the Program monitor instantly shows the picture on camera 2.
Ready three.	The TD sets camera 3 to the preset position.
Take three.	The TD switches to camera 3. Its image now appears on Program.

Figure 8.19. *Live on tape. During a live production, the director calls out commands to his TD, on the right. The TD uses the switcher in front of him to switch cameras and other video sources. The AD, on the director's left, helps with timing and coordination.*

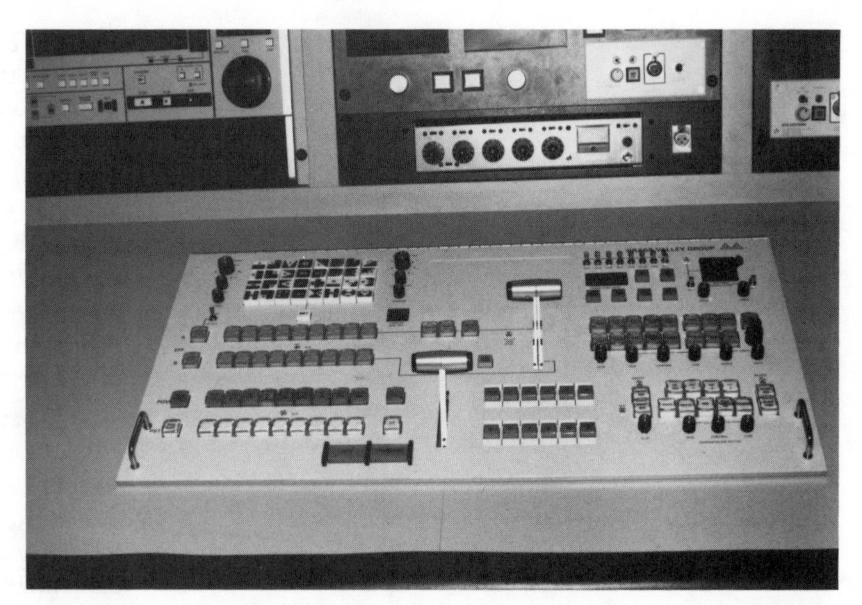

Figure 8.20 *Production switcher. This medium-sized production switcher can manipulate video signals in many ways. Rows of buttons on the left can each provide a different source. The bottom two rows are preset and program buses. The two rows above are effects buses. Squares with black markings above the effects buses provide different wipe patterns. Additional buttons are for chroma keying, mattes, storing effects, transition speeds, and so on.*

and so on until the scene ends or someone makes a mistake, which, of course, is when the familiar "Cut!" is heard.

Just as on the location shoot, the chances of everything going right the first time on this shoot are slim. A camera operator may miss a shot, the TD may switch to the wrong camera, an actor may blow a line, or the director herself may miss a call and have her TD switch to the wrong source.

Chances are a take two will be called for, and the scene will be repeated. Later, when everything has played out to the director's satisfaction, she will call the scene a "buy." The AD will circle that number on her master script, and the crew will move on to the next scene.

The typical corporate studio shoot of this type will be shot in large chunks or scenes, which the editor will later cut together into the final program. As for time, if the action isn't too complicated, the crew can hope to accomplish about fifteen pages of script a day. This means the average corporate script can be shot in a single day in the studio. Because of the setup time and other complications usually inherent in a location shoot, this same amount of shooting on location would probably take two or three days.

Figure 8.21 *A three-camera shoot in action. Cameras 1, 2, and 3 appear, starting from the left. The AD is making a final check to be sure the talent have everything they need.*

9 Audio Production

If you are producing a program with a voice-over narrator or other voices and sounds that do not actually synchronize or "sync up" with specific pictures, some of your production will be audio only.

AUDIO RECORDING IN THE STUDIO

Audio production (Figure 9.1) can be accomplished either on location or in the studio. If the audio is recorded in the studio, it will most likely take place in the area called *production audio.*

The Production Audio Room

Production audio usually consists of two adjoining rooms separated by a glass partition. The first is an *announce* or *narration* booth. This is the soundproof room in which the actor sits when reading his lines. It is equipped with a microphone, a simple seating arrangement, and usually a video monitor on which a program can be viewed as the voice-over recording is taking place.

The second room is the *control booth.* This is the room on the other side of the glass partition in which the audio engineer, the director, and usually the AD are seated. As the name implies, it is from the control booth that the recording session is actually started, stopped, recorded, and played back.

A typical control booth is equipped with a multichannel audio mixing board (eight channels is very common), two sets of speakers (a large, high-quality pair and a small pair that simulate sound as heard through a typical television), a reel-to-reel audio tape recorder, and other sources of sound that can be fed into the system for recording. These might include an audio cassette player, a turntable, and a compact disc player.

Figure 9.1 *Audio production. An audio engineer does a voice-over session by himself. The talent sits in a soundproof room across from him.*

Figure 9.2 *Fingers on the fly. Skilled audio engineers can operate many faders on the fly while at the same time adjusting equalization and other sound modification controls.*

The Recording Session

Before a recording session, the audio engineer usually sets up the room by patching the sources to be used through the audio board on individual channels and routing them into the audio tape recorder. If music and sound effects are being recorded, the sources might be the turntable and the compact disc player. For an actor's narration, the source would simply be the microphone in the announce booth. In some cases, these different pieces of equipment may be *normaled* or preassigned and connected to certain channels on the board.

Assuming the microphone is the source for this session, once the room is set up, the actor is invited into the announce booth. The director and assistant director sit with the engineer in the control booth. Typically, a level check is run by asking the actor to speak into the microphones. As he does so, the engineer checks the volume and tone of his voice passing through the system and sets the equipment to the proper recording levels. At this point, recording can begin.

The director now gives the actor final instructions concerning the type of read he would like. The engineer puts the audio tape recorder into "record," he signals the actor with a hand gesture, and the session is off and running.

Just as in video production, there are usually multiple takes for various reasons. Frequently, the director may want to hear playback of some segment just recorded in order to make his determination. Just as in video production, the assistant director listens intently and takes continual notes on which readings are good and which are bad, why, and which the director buys.

The Audio Script

The script used in audio production may be an unmarked copy of the shooting script for audio use only. Later, when the program moves into postproduction, the editor would use this audio master script to do the necessary voice-over edits and the video master script to edit pictures in under that narration.

Transfer to Videotape

Production audio is usually recorded onto quarter-inch audio tape. Later this is transferred to videotape for use in the off-line and on-line editing sessions. It is best if this transfer is turned in for duplicating along with the video footage shot for the program, in order to keep all the elements of a particular show together. A new producer soon learns how easily audio and video reels can be misplaced when they are not well organized and accounted for.

AUDIO RECORDING ON LOCATION

If audio production for voice-over segments is done on location, it is best recorded wherever the on-camera portions were done. This, of course, assumes that acceptable audio can indeed be recorded at your video location. On location, audio segments are recorded directly onto videotape instead of quarter-inch audio tape. This is accomplished by turning on the camera and either pointing it at the slate for that scene or switching it to the color bar circuit. The VTR is then put into record, and sound is recorded onto the videotape just as any other scene might be, except that the visual part of the picture is not for use.

Sync Source

The reason the camera is required at all is that video recording systems typically require a *sync source* in order to run at exactly the same speed as the pictures being recorded. The camera provides this sync source. If it is not present, the VTR could run at an inconsistent speed, and dialogue may not match the lips of the people speaking.

Field or Studio

Recording voice-over on location is a good idea whenever parts of the script are on and parts are off camera. Using the same microphone placed in the same location on the actor and, if possible, recording in the same general area means that the audio "presence" of individual segments will match well when they are edited together later. This also saves a later transfer session of quarter-inch audio tape to videotape.

On the other hand, if part of the audio for a program is recorded in an audio booth and part in the field, there will be a definite difference in the general quality of those audio segments when they are edited together. Although this presence, as well as ambient background sounds, can be added later in the postproduction process, the booth audio usually will have a noticeably different quality than the audio recorded in the field.

THE IMPORTANCE OF SOUND

One final note on audio recording: Whether it is audio only or audio in sync with video images, sound is often given a back seat to the visual aspect of a program. Many directors and producers have had the "fix it in post" mentality come back to haunt them later in postproduction. The reason is simple: The pictures turn out great, but a sound problem overlooked in the field actually ruins the circled take—and it *cannot* be "fixed in post."

Poor sound can destroy a program just as quickly as poor pictures, poor acting, or poor scripting. The experienced producer understands this, as well as the importance of fighting off the temptation to overlook sound problems in the heat of production. She insists that the proper microphones be used, that competent sound engineers be hired, and that the proper care be taken in production to ensure that the quality of the sound is top-notch, along with the rest of the program.

THE WRAP

Be it location, studio, sound, pictures, or some combination of all four, after several long, hard days all the footage is in the can. The equipment and trucks are checked back in or returned to the rental houses, the set has been struck, and the production is a "a wrap."

Next comes the process of assembling all the elements into an actual program—postproduction.

part four

Postproduction

10 The Elements of Postproduction

OVERVIEW

Postproduction is the process by which the footage shot is edited together into a completed program. The process does not, however, start with editing. It usually starts with duplication of the original footage into a format the editor can work with and preparation of the *master script package.*

Following this step, the editor makes a *rough cut.* This is shown to the clients, and any changes are noted. Next, a *fine cut* is made. The fine cut or a copy of it becomes the duplication master from which the actual viewer copies of the program are struck.

DUPLICATION

Different formats of original footage and different editing systems may create a need for various kinds or combinations of *duplication,* or *duping.* To keep things simple, let's assume that our original footage was shot on three-quarter-inch videotape and will be edited on three-quarter-inch systems. Typical duplication requirements for the footage would then be as follows:

1. *Three-quarter-inch dupes of all the footage:* Often called *B reels,* these are used both as backup copies of the original footage and as a second editing source when certain kinds of edits (to be discussed shortly) are used. These duplicate reels will be used in the on-line editing process only. In our case, during the duplication process, the reels will also have SMPTE Time Code (also to be discussed) added onto channel 3—the *cue* or *address track.*
2. *Three-quarter-inch window dubs of all the footage:* These are work prints of the footage, which will be used in the off-line editing proc-

ess. They have a small window area burned in on the picture in which SMPTE Time Code appears visually as a series of numbers. This time code is identical, *frame for frame,* to that added to the on-line dupes. It is also present on the address track.

THE MASTER SCRIPT PACKAGE

Along with the duplicate footage, a program must include certain written elements before it is ready to be handed to an editor, namely, the *master script package.*

As the name suggests, the primary source of information in the master script package is the master script itself. This is the script (or scripts) the AD marked on all during the production. It will now become an informational tool for the editor, explaining every shot that took place in production and highlighting the "buys" (circled takes) he will need to assemble the show.

Along with the master script should be a shot report, usually noted by the EIC who was working with the VTR. This simply lists the in and out times of the various shots, usually according to a time or footage counter on the VTR itself. The shot report also helps the editor locate certain footage by telling him specifically where the shots are located. A typical shot report would look like Figure 10.1.

Also included in the master script should be paperwork such as listings of any stock footage that is to be used and where it is to be obtained, specifics about any artwork or titles to be used, notes on where to locate music picked by the director, and any other general notes the director or AD may have wanted to pass on.

In short, the master script package should contain any and all written elements, as well as the duplicate footage the editor will need to proceed with his job.

Once these elements have been assembled (usually by the AD immediately following production), the program is ready for editing.

THE EDITING PROCESS AND TIME CODE

In film, the editing process is a manual one; that is, the footage is physically cut and spliced together piece by piece. In videotape, this process is accomplished electronically. If it is done as a first rough assembly of the program, usually *without* the help of an expensive computerized editing system, it is called an *off-line.*

A typical off-line edit bay usually consists of two three-quarter-inch videotape machines (a playback or *source* machine and a *record* ma-

VIDEOTAPE SHOT REPORT

# 76161-3	TITLE THE SUPERVISOR			REEL NO. R-6	DATE 5/18
PRODUCER McPHERSON			DIRECTOR ALBERT	A.D. MARTIN	

FOOTAGE		SCENE NUMBER	TAKE NUMBER	DESCRIPTION
IN	OUT			
001	019	—		BARS AND TONE
019	151	8A	1	JOHN @ TERMINAL -C.U.
	160		2	
	171		3	
	190		(4)	
191	277	9	1	HOST ENTERS @ JOHN'S DESK
	280		2	
	284		3	
	290		4	
	295		5	
	302		(6)	
303	516	9A	1	C.U. HOST
	521		(2)	
522	711	14	1	SUPERVISOR TAKES CALL -W.S.
	900		2	
	016		3	
	121		4	
	209		(5)	
210	327	14A	1	C.U. SUPERVISOR
	461		2	
	608		(3)	
609	800	14B	1	C.U. SUPERVISOR'S HANDS TAKING NOTES
			2	
			3	
			(4)	

Page _____ of _____

Figure 10.1 *Videotape shot report. The shot report gives the editor an accounting of the footage shot plus its approximate location. Counter numbers (on the left) are not the same from machine to machine, but they can get the editor in the general range. In some cases, time code is recorded in the field and used on the shot report instead. In that case, the editor knows exactly where to find the footage he needs.*

chine), several monitors, and a controller. The monitors allow the editor to see the footage he is working with and the time code numbers he needs to perform edits and make an edit list. The controller has a shuttle handle (joystick) and a keyboard, which acts as an interface between the tape machines and monitors, allowing the editor to move quickly through the footage, select edit points, and execute edits very precisely.

Because time code is usually an inherent part of the editing process, let's explore it in more detail.

SMPTE Time Code and Control Track

SMPTE Time Code is a series of digital signals that translate visually into values of time based on a twenty-four-hour clock. The format of these signals has been standardized by the Society of Motion Picture and Television Engineers, or SMPTE, hence the term *SMPTE Time Code.* (SMPTE is pronounced "semptee," but the SMPTE Time Code is often referred to simply as *time code* or just *code.*

Visually, a typical time code number looks like this:

01:05:18:28

The 01 in this code number refers to the videotape hour, the 05 to the minute, and the 18 to the second. The 28 refers to the video frame. (As you may recall, there are 30 of these per second).

This code is usually recorded onto an audio channel or the address track (cue channel) on a videotape using a *time code generator.* On a window dub (remember the work prints we just mentioned) the code is *burned in;* that is, it actually shows up on the screen for the editor.

The edit controller allows the editor to select very specific segments of time on the videotape for transfer onto another tape. For example, suppose an editor wants to transfer a scene that runs a total of 10 seconds, beginning at 01:22:18:10 and ending at 01:22:28:10. In order to do this, he uses the joystick on the controller to shuttle through the footage and stop at two exact points: first, where he wants to begin the edit and, second, where he wants to end it. At each of these points, he enters into the controller the time code numbers appearing in the window—the first as an in point and the second as an out point. Finally, he pushes a button that tells the controller to perform the edit.

At this point, the controller electronically locks in on the specific numbers the editor has chosen. It then synchronizes the record and the playback machines to roll at the same speed. As this takes place, the video and/or audio signals between the two time code numbers selected are transferred *from* the playback tape *to* the record tape.

Control Track

Another, less effective system used for videotape editing is known as *control track*. Control track uses a series of identical pulses (one per frame) recorded on the videotape. While this provides editing control, it does not provide a unique *address* for each frame, a factor that makes the editor's job much quicker and simpler. For purposes of this discussion, we will assume a SMPTE Time Code system is being used.

With this brief explanation of time code, control track, and the basic editing process as a foundation, we can now continue with the actual off-line edit.

11 The Off-Line Edit and Rough-Cut Screening

THE OFF-LINE EDIT

The Editor

If you are not off-lining the program yourself, at this point you will want to hire an editor. Editors, like writers, directors, and crew people, are available on a freelance, per-project basis.

Editors' rates vary depending on the type of equipment they will be working with. An off-line editor who works in the type of edit bay described in Chapter 10 (two videotape machines and a controller) will cost approximately $150 to $300 per day. An on-line editor, who usually works with more sophisticated equipment (to be discussed), will cost approximately $200 to $350 per day.

Off-Line

The off-line edit usually begins after the editor has met and discussed the program with the producer. During this pre-edit meeting, the editor receives any guidelines, preferences, or special instructions he must follow. He also receives his master script package and three-quarter-inch window dubs.

Once he is in his edit bay, the editor's first act will probably be to read the script and get a feel for it from an editing standpoint—whether it is fast-paced, light, dramatic, and so on. At the same time, he will look over the AD's notes to get a feel for the type and amount of footage that was shot to cover the various scenes.

Once he has done this, the editor will view the footage. Sometimes he will log in his own shot list using time code numbers. In many cases, the director will also sit in on this logging session, discussing the footage with the editor and making known her preferences for how certain scenes should be cut.

With these two chores completed, the editor has a very good feel for the script, the producer's and director's preferences, and the footage he has to work with. Now, either with or without the director present, he will begin to edit.

The Editing Process

The editor will start by loading a three-quarter-inch cassette tape that has been *stripped* (prerecorded with time code and black) into the *record* machine. Next she will look at the master script and determine the first shot of the program. Let's assume she finds in the script notes that the first shot is scene 101, that it has been recorded on reel 4, and that take 7 was the buy since it has been circled.

Next, she will load her three-quarter-inch window dub of reel 4 into the *playback* or source machine, refer to the videotape shot report taken by the EIC to find its approximate location, and shuttle through the footage to that spot.

When she has found the slate for scene 101, take 7, she will look over the shot, decide on what she feels is the right in point or place to begin the edit and the right out point or the place to end it. As we have discussed, these in and out points are specified by the time code numbers visible on the screen in the time code window.

At this point, she will also decide on in and out points on her record tape—the tape *onto which* the footage will be recorded.

Figure 11.1 *Off-line editing. The off-line editor checks script notes between edits. Monitors are above her head, the source machine is on her left, and the record machine is just past her head on the right. The small monitor just past her face displays time code plus edit status. The controller is directly in front of her.*

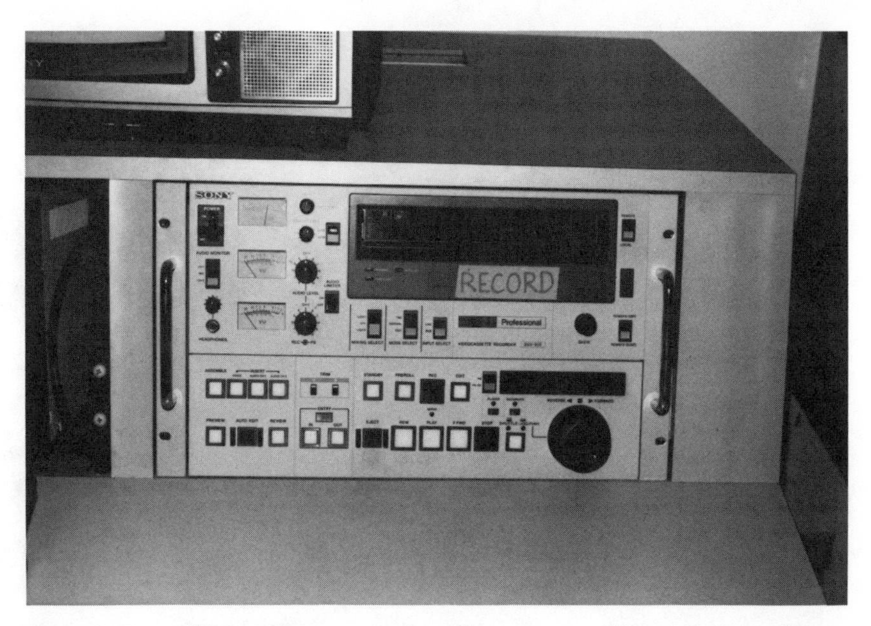

Figure 11.2 *Off-line VTR. A record machine in an off-line edit bay. The machine is put into remote mode, which transfers control of its functions to the controller.*

The editor will now enter these numbers into her controller and probably push the preview button. The preview function lets the editor view what the edit will look like without actually performing it. This is useful because many times the first in and out points she chooses will have to be *trimmed* (adjusted slightly) before she feels the shot is exactly right to edit into the program.

After trimming a few frames at either the head or the tail of the shot, the editor will give the controller the "perform" command, and it will do the following:

1. Roll the record machine to a spot usually five seconds prior to the *exact* time code number on the tape the editor chose as an in point.
2. Roll the playback (source) machine (with the three-quarter-inch window dub footage) to a spot five seconds prior to the exact time code spot on the tape the editor chose as an in point.
3. Once both machines have prerolled and stopped, it will roll both at identical speeds (in sync) and record the piece specified on the source tape onto the record tape.
4. Finally, it will end recording at the exact spot on both tapes the editor selected as out points.

With this done, the editor has laid down her first edit on this program. Now, depending on the system she is working with, she may do a number of things.

Figure 11.3 *Editor controller. This Convergence edit controller is the brain of an off-line edit bay. The keys on the lower left are used to input time code addresses and make adjustments to edits. The joystick, here held in the editor's hand, allows the controller to seize either the source or the record machine and shuttle through footage. Preview and perform buttons are two of the white buttons to the left of the joystick.*

If the system has what is called an *automatic list management function,* the time code numbers she has just chosen will automatically be stored in memory and later recorded onto a computer disk for use in the on-line editing session. In this case, the editor will now be free to go on to her next edit.

If the system she is using does not have such an automatic function, the editor will have to write down those time code numbers on paper. This is called keeping a *manual list,* as shown in Figure 11.4. Whether manual or computerized, this list of time code numbers, often called the *edit decision list,* will later become the backbone of the on-line or final editing session.

Figure 11.4 *Off-line edit decision list. The master edit log gives the on-line editor exact documentation of the shots selected and how they were edited together. This list then becomes the framework for the on-line edit. EVENT is the number of the edit. A/V shows the type of edit. An A in this column means audio only; V means video only; B means both audio and video. SCENE and TAKE are often optional. REEL number must be included. The playback IN and OUT columns list the actual time code addresses of the footage on the source tape, that is, where the footage came* from. *The record IN and OUT columns list the exact addresses on the record tape—where the footage went* to.

MASTER EDIT LOG

401-C	TITLE: Comm. Network Overview				PROGRAM LENGTH: 7 MIN 13 SEC
PRODUCER: McGill		EDITOR: Martin			SHEET 1 OF 3

	EVENT NO.	A/V	SCENE	TAKE	REEL NO.	NOTES	PLAYBACK IN	PLAYBACK OUT	RECORD IN	RECORD OUT
	1	A	104	2	5		05:26:47:16	05:27:21:04	00:01:04:01	00:01:37:19
	2	A	104A	7	5		05:04:10:07	05:04:15:21	00:01:37:17	00:01:43:01
*	3	A2	—	3	1	CORP. MU. CUT 3/2	01:05:35:17	01:06:22:04	00:01:00:00	00:01:46:17
	4	V	105	3	2		02:21:57:26	02:22:04:03	00:01:00:00	00:01:06:07
*	5	V	105	2	2		02:22:29:06	02:22:33:13	00:01:01:29	00:01:06:06
	6	V	105B	1	2	C.U. SCREEN	02:26:19:20	02:26:23:26	00:01:03:23	00:01:07:29
	7	V	105B	4	2		02:24:50:09	02:24:54:29	00:01:06:00	00:01:10:20
	8	V	106	2	2	CAL. MAP	02:22:19:00	02:22:24:24	00:01:07:19	00:01:13:13
	9	V	106D	6	2	FLORIDA MAP	02:26:17:15	02:26:21:11	00:01:03:23	00:01:07:19
	10	V	106E	1	2	OAHU	02:28:54:18	02:29:03:05	00:01:11:13	00:01:20:00
	11	V	107	1	2		02:27:49:26	02:27:54:05	00:01:15:09	00:01:19:18
	12	V	107A	9	2		02:24:49:25	02:24:55:21	00:01:19:04	00:01:25:00
*	13	V	107	1	2	SERVICE ORDERS	02:24:49:25	02:24:56:02	00:01:17:12	00:01:23:19
*	14	V	107E	1	2		02:23:21:14	02:23:27:02	00:01:21:01	00:01:26:19
	15	V	108	3	3		03:11:25:17	03:11:29:29	00:01:26:19	00:01:31:01
*	16	V	109	2	3		03:12:11:01	03:12:16:24	00:01:28:27	00:01:34:20
	17	V	110	2	3		03:09:33:12	03:09:36:11	00:01:34:20	00:01:37:19
*	18	V	110A	3	2		02:44:03:23	02:44:10:14	00:01:37:19	00:01:44:10
	19	B	110	2	2	INTERVIEW BITE	02:44:10:14	02:44:24:16	00:01:44:10	00:01:58:12
	20	V	111	1	2		02:23:22:09	02:23:27:27	00:01:21:01	00:01:26:19
	21	V	111A	4	3		03:11:22:00	03:11:33:00	00:01:26:19	00:01:37:19
	22	A	111B	3	5		05:05:23:00	05:05:30:26	00:01:49:07	00:01:57:03
	23	V	112	3	3		03:52:23:24	03:52:29:19	00:01:51:15	00:01:57:10
	24	A	113	1	5		05:05:50:00	05:05:53:00	00:01:56:06	00:01:59:06
	25	A	113A	1	5		05:06:21:19	05:06:33:00	00:01:57:10	00:02:08:21
	26	V	114	2	1		01:04:37:07	01:04:42:12	00:01:57:10	00:02:02:15
	27	V	114	2	1		01:04:37:07	01:04:48:15	00:01:57:10	00:02:08:18

Now, working within the parameters of this system and according to the desires of the producer and director, the editor will move through the program repeating what she has just done—selecting shots using the master script and log sheets; entering the specific time code numbers of the ins and outs of those shots into the controller; and previewing, trimming, and finally performing edit after edit until her off-line or rough-cut version of the program is complete.

At this point, she will return to the producer the original package containing window dubs and master script package, with two additional elements: a completed rough cut of the program and either a manual (paper) list of all her edits or a computer disk with the numbers stored on it.

Now, with the rough cut complete, it's time to call the client.

THE ROUGH-CUT SCREENING

If your client has never been involved with the production of a videotape program, she will probably be waiting impatiently for a first look at what you have developed. That first look normally comes in the rough-cut screening.

It is very important to make sure the client understands that the program, at this point, is exactly what the name implies—a *rough* cut. If, after weeks of anticipation, the client comes to the rough-cut screening expecting music, dissolves, and special effects—a perfectly polished program—she will probably be in for a disappointment.

Another equally important point to clarify is that this is the client's last real chance to make changes in the program. If this is not made clear, you could be in for some headaches and, as we will see, some added on-line expenses.

To keep from running into these problems, it is a good idea to open each rough-cut screening with an informative introduction, which might go something like this:

> *What you're about to see is what we call a rough cut. Basically, that means a rough assembly of your program, shot by shot. What it* doesn't *include is the music, titles, any special effects we may add, certain basic effects like dissolves, and the general fine tuning and polishing that come in the on-line—the final editing process.*
>
> *Today's screening, then, isn't so much for aesthetic reasons but more for content approval. Is all the information there? Is it placed in the right order? Has it been visualized correctly? And so on.*
>
> *Once we've established these things or noted any changes we need to make to accomplish them, we can then move the program into on-line for those final touches I mentioned.*

Another key point to remember is this: Today is basically the last *approval point you'll have before I deliver your final product. After today, it usually becomes very expensive and complicated to make even little changes. So whatever questions or concerns you may have, by all means bring them up in this meeting.*

At this point you might explain to the clients that they will be seeing the time code window and numbers at the bottom of the screen, and that these will be gone in the final version. Then, if there are no questions, you can proceed to show the program.

Most new clients will not even notice the elements you have told them are missing. They will probably be so delighted to see their message actually playing on television that they will be unaware of whether a cut should have been a dissolve or whether a montage has music.

In either case, however—delight or disappointment—it is important to pinpoint any specifics that need to be added, deleted, or changed in any way before the program moves ahead. It is equally important to take very specific notes on these changes. That is because the warning that this is the last approval point is basically true. If the program needs more changes after it has been on-lined, depending on the type of system that was used, the cost and complication factors can multiply dramatically.

ADDITIONAL NOTES

Here are a few other things to note about rough-cut screenings.

It can be very beneficial to have the editor or the director attend this meeting, if time and money allow. These two, especially editor, are much more familiar with the footage than you are by now.

If the client is concerned about a certain shot, for instance, it is the editor who will know if there was another take or another shot that might substitute. The director, who planned the visualization very carefully and, in many cases, got extra coverage of certain scenes as backup, should also be able to help in situations like this.

After this meeting, you will have to decide whether to send the program on to on-line or hold it back for more off-line work. That decision is usually based on the types and amount of changes that have been decided on. Another factor in the decision will be whether or not you have your own on-line facilities or must rent them.

The rule of thumb is to make essentially all changes in off-line editing, because a typical on-line system rented from an outside facility can cost $300 to $600 per hour.

If you own the on-line equipment, you have a little more flexibility because you will not be paying these fees. In either case, however, if you do not clean up all the changes in the off-line, you should at least resolve most of them. This will ensure that the on-line session moves smoothly and cost-effectively.

12 The On-Line Edit and Audio Sweetening

THE ON-LINE EDIT

Earlier, we established that the term *on-line* refers to the final editing process. In a few cases, the on-line is accomplished on a system as simple as the one used in off-line editing, sometimes even the same system. Much more often, however, an on-line takes place *on-line,* with a highly sophisticated, computerized editing system, as shown in Figures 12.1 though 12.4.

All the various types of on-line editing systems do basically the same thing: They control multiple videotape machines; interface with switchers, special-effects generators, audio mixers, and other equipment; and, in doing so, assemble the final program from original footage (*not* work prints) and usually first-generation duplicates (B reels) of the originals.

In the off-line edit, as you will recall, a rough cut was made from duplicate footage with a time code window superimposed. Because there was only a single playback or source machine in the off-line edit bay, the system was capable of cutting together shots from only one source. This is called *cuts-only* editing. During that rough cut, the editor, either manually or via the system itself, developed a list of time code numbers that corresponded exactly to each edit event.

These time code numbers correspond *exactly* to the same pictures on the original footage—not in the form of a visible window but, rather, as a series of digital signals on the address track. This, as you may recall, was accomplished during the initial duping after production.

Using these numbers for reference and control, then, a sophisticated on-line system can control more than one playback or source machine and is capable of using any number of special effects, depending on the interface equipment associated with it. Further, it can perform edits *automatically* once it has the list of time code numbers entered into its processor. In an ideal situation, then, you could get an on-line edit session that proceeds as follows:

1. The edit decision list from the rough cut is fed into the on-line machine and stored. This is accomplished by disk input if the au-

Figure 12.1 *On-line edit suite. An on-line session in progress. The keyboard and monitor are in front of the editor. The audio board is past him on the left. In the foreground on the right is a video switcher, and behind it is a keyboard for a character generator. The computer processor is in an adjacent room.*

Figure 12.2 *Keyboard for an on-line editing system. The keyboard places many sophisticated functions at the fingertips of the skilled editor.*

Figure 12.3 *A typical on-line editing system: keyboard and monitor surrounded by accessory equipment.*

Figure 12.4 *On-line editor loading a one-inch tape onto a VTR. This and other VTRs will then be controlled by the on-line editing system.*

tomatic list management system was used, or, if the list is manual, by entering the numbers on the system's keyboard.

2. Anywhere from two to four, five, or even more original cassettes are then put in individual videotape machines connected to the on-line editing system.
3. The editor gives the system a series of commands that tell it to "auto assemble."
4. Then, as she looks on, the on-line system synchronizes and rolls the proper machines at the proper times to edit the entire program together automatically before her eyes.

This is an ideal situation. In the real world, it rarely happens this way. It is much more likely that chunks of the show will be auto-assembled, whereas other parts will be assembled manually, edit by edit. The amount of editing that can be done in the auto-assemble mode is determined by how closely the time code list developed by the off-line editor reflects the final version of the program. In other words, if the off-line of the program reflects *exactly* how the final program should look, and the editor's list accurately shows each of the edits performed, then the on-line session could be done strictly in the auto-assemble mode. But if the rough cut is *close* but not in exact agreement with the written list, the changes required will negate the auto-assemble process.

Figure 12.5 *Small video switcher. This fairly small video switcher is still a very powerful tool in production or postproduction. This one is part of an on-line edit suite.*

Figure 12.6 *Character generation. The character generator operator developes electronic titles for a program. The system is much like a very powerful word processor. Various fonts are available, and the titles developed can be made to roll, crawl, or reveal, line by line, in the program.*

Obviously, then, from a producer's point of view, it pays to have the program as close to the final version as possible before taking it into the on-line session—*especially* if you are renting the on-line facility.

SPECIAL EFFECTS

It is also during the on-line edit that special effects—dissolves, wipes, and so on—will be put into the program. Accomplishing these effects requires the use of more than one playback machine because two sources are being mixed visually during one edit.

A *dissolve,* for example, is made up of two pictures—one fading out and the other fading in—at the same time. Both these sources, often called A/B rolls, must be recorded at the same time from two different source machines onto a third one—the record machine. The on-line system is designed to accomplish this with its expanded machine control capabilities and its use of the video switcher.

A *wipe* is a similar transitional effect in which, usually, a line crosses the screen, bringing in a new picture as it passes. A wipe does not require fades, but it does involve two pictures at once—one already on the screen as the other wipes it off. For this reason, a wipe also requires two source machines and one playback machine, as well as the switcher.

Digital video effects (DVE) are also normally an on-line function (see Figure 12.7). These are accomplished with special pieces of equipment, which digitize the video image and then manipulate it in any number of ways. The picture can be shrunk, expanded, flipped away off the screen, posterized, and so on. DVE, used in conjunction with the switcher, adds even more capabilities to the system.

Superimposed titles will now be added, also. Usually, these are developed on a character generator, either in the on-line edit bay or nearby. The titles will be keyed in or "supered" over the proper pictures, again using the video switcher.

When the program is totally complete visually, the final step in the postproduction process is often to "sweeten" its audio track.

AUDIO SWEETENING

Audio *sweetening* refers to enhancing a program's sound. When the requirements are simple, this can often be accomplished in the on-line session, working with the audio channels available on the existing vid-

Figure 12.7 *Digital video effects (DVE). The digital effects generator picks up image manipulation where a typical switcher leaves off. The video signal is digitized. This allows it to be shrunk, expanded, posterized, exploded, disintegrated, flipped, and spun—to name just a few possibilities.*

Figure 12.8 *On-line audio board. Small audio mixer typical of an on-line edit bay. The audio signals from various VTRs are set up to feed through the mixer. This gives the editor individual control over each channel.*

eotape. Typically, channel 1 would contain all dialogue or narration. Channel 2 would contain any music and sound effects required. Channel 3, the cue or address track, would contain time code.

If additional sound effects are required and time permits, the audio mixing board in the on-line room may be used to mix, say, channels 1 and 2 together onto *only* channel 1. This would free up channel 2.

In an on-line editing room, however, mixing sound can become quite expensive at rates of $300 to $600 per hour, especially when the mixing requires more than a simple addition or change. When such requirements exist, then, by virtue of either volume or complexity, an audio sweetening session is normally arranged. It takes place in a room dedicated strictly to sound.

The Sweetening Room

The sweetening room normally contains an audio board with a number of individual audio inputs and a multichannel audio recorder. Also present are a television monitor and a videotape machine with three-quarter-inch playback and shuttle (visual search) capabilities. Finally, the room contains various sources, such as turntables, cassette recorders, and disk players, and assorted pieces of equipment used to modify the sounds in the program in different ways. These might include filters, compressors, reverberation units, and so on.

Figure 12.9 *The mix. The producer and audio mixer prepare for a mix. The audio board is a twenty-four channel unit. A reel-to-reel tape recorder and turntable can be seen in the foreground. A sixteen-channel audio recorder (which, after lay-down, will contain special time-coded audio tape) is at the upper left. The video monitor is used to watch the program for placement of sound effects and music.*

Figure 12.10 *The mixing board. A multichannel board like this can seem intimidating to the inexperienced onlooker. The knobs, buttons, and faders on each vertical row, however, do exactly the same thing—each to a different audio channel.*

Figure 12.11 *Three-quarter-inch control. A three-quarter-inch VTR used in a sweetening room. The shuttle controller at the lower right allows the operator to move through the program backward or forward. The audio recorder is interlocked, or slaved, to follow. Sound effects and music can thus be placed to sync with the pictures.*

The Laydown, Mix, and Layback

The first step in an audio sweetening session is to *lay down* the existing audio channels. Channels 1 and 2 from the on-line edited version of the program—the edit master—are laid down off of the original videotape onto a multichannel audio tape, which may have up to sixteen (or more) separate channels. Time code is also recorded onto this tape on one of those channels. Simultaneously, a three-quarter-inch videotape copy of the entire program with *identical* time code in the form of a window dub is also made.

With both these items in hand, the audio engineer and usually the director or producer now head for the sweetening room. There, the audio tape is placed on a special audio recorder, which, like the tape itself, has multichannel recording capability. The three-quarter-inch window dub is placed into the videotape machine with shuttle capability and is viewed on a monitor.

Normally, the multiple-input mixing board is normaled or prepatched so that each of the sixteen channels from the audio recorder is routed into a separate board input. This allows individual control over each channel with regard to volume, filtering, equalization, and so on. Once any necessary patching of sources has been done, reference levels are set to ensure that the volumes remain consistent on all channels.

Figure 12.12 *Audio recorder. A sixteen-channel audio recorder loaded with tape and ready to operate. Multiple volume unit (VU) meters allow the operator to view each individual channel as it records or plays back.*

Figure 12.13 *Audio board. An engineer's view of the audio board. VU meters monitor each channel.*

At this point, the room is set. Now, because the audio tape and the three-quarter-inch window dub have identical time code and are *interlocked* (able to roll in perfect sync) to the same shuttle control, the audio recorder is *slaved* to follow the three-quarter-inch visual copy of the program. This allows the engineer to shuttle back and forth through the program, finding places for music and sound effects that will match the pictures perfectly. These effects are recorded onto the audio tape on separate channels in a series of *passes.* Each may be in perfect sync with some visual part of the program on the three-quarter-inch window dub.

Once all the effects recording has been done, the actual mix takes place. Using the mixing board skillfully, the engineer begins combining all the channels on which he has recorded, adjusting various volume, tone, and filtering characteristics as he goes. In the end, the entire mix, consisting of perhaps twenty individual channels, is combined onto one or two single channels.

As a final step, this mixed audio recording is now rerecorded, or *laid back,* onto the original edit master of the program from which it was originally removed. Because all this was done in sync with the time-coded three-quarter-inch window dub, all sounds remain in perfect sync with pictures on the edit master itself.

To summarize briefly, then, an audio sweetening session is accomplished as follows:

1. The existing audio channels from the edit master of the program (the final edited copy) are *laid down* simultaneously onto multi-channel audio tape and a three-quarter-inch window dub, both with identical time code.
2. In the sweetening room, various sound sources (music, background ambience, etc.) are added to the audio tape on different channels and modified as required. Each can be placed in its proper position on the audio tape using the three-quarter-inch window dub as a picture reference.
3. The multiple audio channels are *mixed down* to one or two final audio channels.
4. The final mixed audio channels are *laid back* onto the original edit master of the program in perfect sync with the pictures.

The result is a completed *dupe* master videotape with a sweetened sound track, from which duplicate copies can now be made in any format and sent to the client for use in the field.

part five

Evaluation and Client Relations

13 Why Evaluate?

Many producers consider evaluation one of those tedious administrative chores that they must do whether they like it or not. Most of the time, they don't like it; and that's understandable, for two reasons. First, evaluations mean more work—paperwork, fieldwork, phone work, and mental work—on top of an already overloaded schedule. Second, evaluations typically are not considered part of the production process. They do not involve writing, producing, or directing—the things a producer tends to see as his purpose in corporate life.

As with many of the other elements covered in this book, the producer may be able to hire someone to do her evaluations, or there may be a departmental secretary or evaluations specialist. In that case, there may be no resentment, because someone else will do most of the work. If there is no one else, however, and the producer wants to rid herself of the feelings of annoyance that accompany any evaluation assignment, she simply needs to ask: Why evaluate?

THE EVALUATION PAYBACK

Evaluating programs pays back two important elements to any corporate television department. The first is obvious—feedback on the effectiveness and audience acceptance of the work being distributed. The second may be less obvious, especially if the producer has not been in the corporate television business very long. Media managers or administrators might call it *documentation,* but a better word is probably *ammunition.* That's because favorable evaluations provide proof from the people on the line that the programs being produced are *worth the company's expense.*

Let's consider these two items one at a time.

Feedback and the Need to Improve

The employees in any company are to the corporate television department what the general public is to the producers of any evening's broadcast programming—the single most important ingredient in their survival. If the audience isn't happy, they're not going to watch. And if the audience doesn't watch . . . well, we all know what happens then.

By evaluating what the audience likes and dislikes, then, and acting on those evaluations, the corporate television department is able continually to improve its product and its image and thus remain valuable to the company's employees.

This latter statement might seem to imply that once your television department becomes very good at what it does, the need to evaluate goes away. No such luck. I have yet to meet a corporate television producer who feels he or she cannot improve on the quality of the product. This is because the nature of television work and the technology it uses are constantly changing and improving. Writers and directors are continually looking for a better theme or a more interesting shot. The directors of photography are continually striving for just the right contrast ratio and lighting mood; engineers are continually looking for higher resolution or an ingenious way to transmit or receive a video signal.

The equipment with which all of these people make a living is also constantly improving. Cameras, recorders, editing systems, and sound systems are continually getting smaller, lighter, more sophisticated, and capable of delivering better quality.

Achieving constant improvement, then, becomes a way of life for the corporate television department that wants to stay in business. And evaluations are the quickest and easiest way to find out exactly what needs improving.

But audience feedback, as important as it is, is still only half of the big picture.

Documentation: Ammunition against Executive Attack

Although corporate television departments are a rapidly growing part of the corporate world, they are also a very vulnerable part, for two simple reasons. First, executives who know little or nothing about what the corporate television department really does tend to see it as a luxury that can be pared down or even eliminated at budget-cutting time. Second, it can be very hard to disprove what these executives believe because it is difficult to prove just how much help the television department is to the bottom line.

For example, how much money would an excellent program on how to use a new test meter actually save company X in a year if all 100 craft employees who will use the meter see the program? Is there a

case study
Does a Videotape Pay Back Its Cost to the Company?

As I have noted, proving the value of a videotape can be very difficult for the television department. Once, however, I had the pleasure of being involved with a program on which the payoff could not be disputed.

I was first approached by the security department to produce a program on drug abuse in conjuction with the medical department. The plan was for both departments to contribute support and expertise. The result would be a single program, aimed at all employees, exploring the terrible results of drug and alcohol abuse.

Shortly after the writing process got underway, it became apparent that the two departments did not see eye to eye. The security department wanted to produce a simple ultimatum: "Use drugs or alcohol on the job and you're not only out the door, we'll also help send you to jail."

The medical department, on the other hand, understood that drug and alcohol problems are not always so cut and dried. They (and I) wanted to do a program that would be a frank exploration of drug and alcohol addiction. The program would also send the message that the company was there to help any employee who showed that he or she really wanted to break an addiction and start a new life.

After a few months of departmental bickering, the program was made and distributed to all employees. Shortly after this, our department was undergoing a particularly rigorous period of grilling by the budget people. Finally, in an attempt to gather support from our clients, my boss came up with an idea. He asked that each producer get on the phone and see if any clients had information on ways our programs had paid for themselves in concrete terms.

One of my calls was to my client in the medical department who had helped fight to keep the drug abuse program going.

"Concrete terms?" she said.

"Right," I responded. "Dollars and cents."

"I can't offer that kind of 'concrete,'" she said, "but how about this: Since our drug abuse program has been out, over twenty people have called our Employee Assistance Hotline as a result of watching the videotape."

I was ecstatic. Twenty people! How could anyone possibly measure the cost of twenty lives? Yet how could anyone ignore the value of a program that had helped save them!

I told my boss immediately. First his jaw dropped, then he went to the phone and called the budget committee. For the rest of that year, they left us alone.

way to calculate how much more time it would have taken these employees to do their jobs or how much rework might have resulted if, instead of viewing a videotape, they had simply been given the user's manual and told to go ahead and use the meter? Unfortunately, such benefits are difficult, if not impossible, to quantify. Provided the television department really *is* helping out, however, properly designed and executed evaluations can add up to an impressive showing. With that background in mind, let's first consider a few of the basics and then discuss two of the many types of evaluations you might choose.

WHICH EVALUATION?

Evaluations can be formal or informal, written or verbal. They can be pre or post evaluations. They can be given in person or sent in the company main. But whatever type you choose, a good evaluation should accomplish several things. It should:

1. Tell the producer if a program accomplished the objectives originally established in the PNA.
2. Tell the producer if those objectives were the right ones to solve the problem on which the program was focused.
3. Tell the producer how acceptable the program was aesthetically to its audience.
4. Give the producer input on ways to improve those areas that fell short of the audience's expectations.
5. Be simple enough not to intimidate either the audience or the producer who is developing and administering it.

As an example, you may remember that in the section on writing and design, we used a PNA dealing with customer service as an example. The objectives in that PNA were as follows.

Having viewed the proposed program, the audience would be able to:

1. List the following three proven customer-handling techniques:
    a. Express genuine concern for the customer's needs or problems.
    b. Make every effort to accomplish what the customer wants, or follow up as needed.
    c. Let the angry customer vent his or her frustrations without interruption.
2. State the main reason that proper customer-handling techniques are personally beneficial to them.
    a. Handling customers properly means less frustration and therefore greater job satisfaction.

A written evaluation of this program that would satisfy the five criteria just stated might look like Figure 13.1. Obviously, this evaluation uses various formats—true or false, multiple-choice, and essay responses. All however, are focused on one objective: revealing whether the program was effective and, if not, what could have been done to make it effective.

The first three questions, for instance, are extremely important. Employee opinions on what is a good or bad use of their time, whether the characters were believable or not, and whether the program was helpful to them are key general indicators of its credibility.

The next two questions explore whether the program actually taught the audience what it was supposed to. Even if they indicated they had a great time watching the program, the question remains: Did they *learn* anything?

Question 6 asks for a frank "will/will not" response. If the preceding answers were generally positive and this one is "will not," something is definitely wrong.

Finally, the essay question allows the employee to express his or her feelings about the program in whatever words he or she chooses. At the same time, it encourages employees to offer suggestions or make statements about anything you might have missed in designing the evaluation.

ADMINISTERING THE EVALUATION

The actual administration of an evaluation can be handled in a number of ways. For example, 100 forms like the one just discussed can be mailed with the videotape to the person who will present the program. He would then pass out the evaluations after the audience had viewed the program, gather up the completed forms, and send them to you by company mail.

Although this is certainly one valid way to accomplish an evaluation, time permitting, the best way is to go to the field and do it yourself. There are two main reasons for this. It allows the producer not only to explore the results of her work on paper but also to see it in the live reactions of the people it was intended to help—her prime audience. This can be a very rewarding or, sometimes, a very embarrassing experience. Further, a field visit usually allows a little time for informal discussion in addition to the written evaluation. This, too, can be an important eye opener, as some employees may be better able to express themselves verbally than in writing.

What Next?

Once your evaluation is completed, there are many options, but in one way or another most of them boil down to this: You tally the results,

PROGRAM EVALUATION

"Professional Customer Service -- How It's Done"

In order to evaluate the program you've just seen and improve our future programs we would like your frank, thoughtful opinions and answers to the following questions.

	TRUE	FALSE
1. In general, watching this program was a good use of my time.	____	____
2. The characters in this program were believable.	____	____
3. The techniques this program taught will help me to do a better job when dealing with customers.	____	____

4. According to this program three effective techniques a Customer Representative can use when dealing with an irate customer are:

1. ____________________

2. ____________________

3. ____________________

Figure 13.1 *Sample program evaluation questionnaire.*

5. Handling customers in a professional manner can benefit me by taking some of the ________________ out of my job.

6. I WILL/WILL NOT use the techniques shown in this program.

   (Circle one response. If your answer is "will not" please explain below.)

   ____

   ____

   ____

7. What could have been done to improve this program? Or, do you have ideas for other programs which might be helpful to you in your job?

   (Please use the reverse side if more space is needed.)

   ____

   ____

   ____

   ____

Thank you.

Figure 13.1 *(cont.)*

look for the trends that seem to indicate either very good or very bad opinions from a substantial number of viewers, and then take action—change something in your production process to improve things.

In addition, be sure to document your results as ammunition for that day in the future (and it is likely to come eventually) when some executive wants to know if the "television people" are really doing the company any good.

Indices Evaluations

The evaluation outlined here is only one of many types. Another type worth exploring is what is called an *indices evaluation.* An indices evaluation looks not so much at what the audience felt or learned from the tape but, rather, at what the program actually did to the bottom line.

An indices evaluation might be applicable to the back injury videotape discussed in Chapter 2. You may recall that in that case, the client said the company had lost $90,000 as a result of back injuries in the preceding year. Obviously, then, this safety representative is documenting the number of back injuries occurring and their cost. If this is the case, one of the objectives for that videotape might have read:

1. Provided this program is properly administered to all employees, a 20 percent savings in back injury expenses will be realized by the company during the year following its distribution.

This would be evaluated simply by looking at the records or *indices* kept by the safety department for the following year, documenting the results, and comparing them with that $90,000 figure from the previous year.

Considerations

There are three considerations to keep in mind when doing this type of evaluation. First, since the results tend to be longer-term than audience evaluations, they can require more persistent follow-up after the program is long out of sight and out of mind. Second, indices evaluations are riskier because there is no gray area: The program either does or does not affect the bottom line, and the results are there in black and white for everyone to see. Finally, such evaluations tend to be extremely credible to those executives who are apparently oblivious to everything but the bottom line.

For this last reason alone, if a producer is confident that her work is really making a difference, then it is well worth her time and effort to do this type of evaluation.

A FINAL NOTE ON EVALUATIONS

A sampling of only about 1 or 2 percent of your entire audience will usually give you an accurate evaluation, provided you evaluate a random cross-section. If 1,000 company employees will see a program, and you manage to gather 20 of them and do an evaulation, your responses should reflect accurately the reactions of all 1,000.

With this in mind, then, we can ask once more: Why evaluate? And now we have an answer: Because evaluations are crucial to the growth and survival of any corporate television department.

14 Future Survivors

TODAY AND TOMORROW

Corporate television is a rapidly growing field. Many very large organizations, realizing the power of television, are beginning to find their own creative uses for it.

Corporate Broadcast

One very good example is the use of corporate broadcast networks. This involves linking a number of company locations, usually with the use of microwave transmitters and receivers and/or satellite links. It virtually eliminates the need to duplicate programs and distribute cassettes because the television studio can simply *broadcast live* to each company location.

To appreciate the value to a company of this type of broadcasting, imagine the CEO who wants to get an urgent message out to the entire company immediately. If he has a regular television unit, he will probably call a crew to his office to record his message. It then has to go through the postproduction process described in this book, which at best will probably require an overnight session. After this, the tapes must be duplicated and sent through company mail, probably at least a two-day process, or hand carried to each location.

His other choices would be either to spread the word by telephone, which means the message will be just a bit different for each person down the line, or by sending a letter or memo. Both of these approaches would get the job done, but they are not the most effective ways to handle this type of communication.

If the CEO is able to broadcast, however, he simply calls the studio and makes arrangements to go live whenever he wants. When the broadcast goes on the air, employees at company locations simply stop work and view the message on a television set somewhere in the room. It's much quicker, more personal, and more effective than a telephoned or written message. In fact, such a broadcast could be equaled only by in-person field visits to each work location.

Companies like J. C. Penney, GTE, Hewlett-Packard, Ford, Federal Express, and Merrill Lynch, to name a few, are currently using such broadcasts to communicate to company locations on a national scale.

Traditional Video Units

The use of traditional video units is also on the rise. Although recording and editing may take a little longer than producing a live broadcast, many executives realize the value of using standard cassettes and regular VCR locations throughout the company.

Corporate Television: A Sure Thing?

With these promising trends evolving daily in the corporate world, the new producer might begin to think of his video unit as a sure thing—a service viewed by top executives as indispensable, even in tough economic times. Unfortunately, however, that is rarely, if ever, the case. Many executives still consider the words *corporate television* synonymous with "fat," "a luxury," "overhead," and so on. And the truth is that, in *some* cases, they are right.

It depends heavily on the people in the television unit themselves whether the department will prove to be a survivor when the corporate belt needs tightening. More specifically, there are three things television people can do to help ensure their own survival:

1. Produce high-quality programs at a cost-effective price.
2. Maintain strict audience loyalty.
3. Remember that the client—the employee who is requesting their services—truly holds the key to the department's survival or its demise.

PRODUCING HIGH-QUALITY PROGRAMS

The word *quality* is what we have been discussing throughout this book. It is synonymous with *survival* in any business, and it cannot be overemphasized. The television department that survives is the one that never rests on its laurels or on the success of its last program. Instead, its producers are continually looking for new and creative ways to provide the company with training, motivation, and information.

The successful department's creative people—the writers, producers, and directors—are never satisfied. They are always sure that the most effective program they have ever made is the next one up on the schedule—no matter how good the last one was. Finally, they gauge the quality of a program not by how many awards it has won but, rather, by how much help it has been to the employees for whom it was made.

Awards are certainly a valid and generally well-deserved form of recognition. It is true, nevertheless, that some very expensive, award-winning corporate programs have been flops with the audiences for whom they were made. In business terms, these programs were simply a waste of money. Awards don't help companies pay their bills and corporate executives who are constantly scrutinizing the bottom line are the first to recognize this.

MAINTAINING AUDIENCE LOYALTY

What might make the people in a corporate television department abandon their audience loyalty? The answer is simple: the television process itself.

It takes creative people to make good television programs. But these same creative people sometimes get so wrapped up in television itself that they forget who is buttering their bread, so to speak. They use special effects that happened to look "neat" on "Wide World of Sports"; or they write a humorous script, not because it's the right way to reach an audience, but just to be funny; or they light a scene in a certain way because they saw it in a recent film and it was "dramatic."

Creativity is an asset to any television program—when it is properly guided. A series of ultraslick programs, however, mixed with a stack of poor evaluation results—or, for that matter, no evaluations at all—are a sure sign that the television department is headed for self-destruction.

What's the solution? Think in terms of the audience instead of how exciting a special effect might be. Ask yourself what really matters to the office clerk, the welder, and the marketing manager. Pose the question: If I do it this way, even if it looks fantastic, will the audience *really* accept it? Will it *really* have the desired effect on them? Finally, and most important, is it worth the time and money?

The producer who survives learns to ask himself this question often, particularly during the script and preproduction stages. He also finds that defending the decision to change program elements because of audience considerations can sometimes seem like a full-time job because some writers, directors, and editors are so wrapped up in the television process that they do not understand the issue of audience loyalty. The need to guide this creative pressure is one of the main reasons the position of producer is a necessary one on any production.

case study

Awards versus Audience Acceptance

A new producer friend of mine once had an assignment to produce a program that would show how a specialized work function would be done five years in the future, after it had been totally computerized.

I remember a telephone conversation we had when the project was beginning. As we talked, he made the point that it was extremely important that the production be a good one, because his company would be implementing this computerized system soon, and employee reaction to it so far was extremely negative. He went on to say he had a fairly large budget for the program and wanted to produce something different—"a real attention-getter."

The producer brought in an expensive writer and began hashing out treatment ideas. After several attempts the writer came up with what my friend felt was a winner. The program would feature two current day employees (actually actors) who, while carrying out the laborious present-day functions of their jobs, would suddenly be transported into the future. There they would be taken on a magical journey into the brain of one of the computers, where they would learn all the wonderful features of the new system. They would return from their experience expounding the virtues of the new system to fascinated co-workers, who would all eagerly await the changeover.

In order to record this "journey" realistically on videotape, several elaborate sets would be built in miniature and the actors would be chroma keyed into the scene. *Chroma keying* is a method of shooting actors against a background of one color, usually blue, and shooting the sets they are to appear in separately. The two pictures are then combined electronically. For this to be done successfully, great care must be taken in working out camera moves, the movements of the actors, and lighting. Needless to say, this all adds up to time and money.

The program was eventually produced at more than double my friend's original (already very large) budget of $60,000. It was nearly a year in the making, which in corporate television is a very long time.

When the program was completed, the producer and everyone in his media department were delighted with it—so delighted, in fact, that before its release they submitted it to a television competition, where it won first prize. The creativity of its direction and writing were praised.

When the program was actually distributed to employees, however, their responses, received on the accompanying evaluations, were disastrous. Employees viewed the program as a silly, insulting attempt to convince them to change their minds about something they viewed—now more than ever—as a fiasco.

The reason was simple: The producer, the writer, and the director had sold out audience loyalty for the sake of production value itself. They produced the program based solely on what *they* felt was slick and creative, paying no attention at all to what the *audience* would accept.

The lesson? New creative ideas should always be considered and, when appropriate, used—but *never* without paying close attention to the target audience.

MAINTAINING GOOD CLIENT RELATIONS

I once had a heated conversation with a videotape executive producer about whether or not a proposed training program would be effective with its intended audience.

The executive producer's position was that the client wanted the program to be all nuts and bolts—footage of actual work functions as they happened on the job. The client's concern was that he get as much bang for the buck as possible out of the program and get his new employees productive as quickly as possible. The executive producer reasoned that in order to get funding for the project, he should give the client exactly what he wanted.

My position was a little different. I felt that since the audience was made up of *new* employees, simply showing a series of job functions would not be totally effective. The audience, I reasoned, would first need an *overview*—a nonfunctional look at why certain things were done, how they interrelated with other job functions, and how all these elements fit together to form the big picture.

After some discussion, the executive producer said: "Consider the project something like dog food. Granted, a certain kind may be better for the dog itself. It's not Fido that writes out the checks, though, it's his master. And what the master wants is what *he* thinks is good for his dog. . . . Get it?"

"I get it," I said. "But the master isn't trained in the content of dog food or videotape. He doesn't really *know* what's best for his dog. What I want to do is make him aware of that. Then I want to give him the kind of dog food that will make his dog healthier than ever."

This conversation may sound silly, but it actually did take place, and it points up a dilemma that continually faces corporate producers—how to keep the client happy, even when what he wants is wrong.

The answer most often lies in the confidence that comes with experience. A new producer would probably find it very difficult to argue her point with a client, especially a high-level executive, even though she might feel she was right. On the other hand, a producer who has shaped, screened, and evaluated many programs in the past would feel much more confident in trying to persuade the client to see things her way. Whether you are new or experienced, however, the right thing to do is to make your case as eloquently as possible. Remember, it is for the client's own good.

Here are a few points to remember that may help in that conversation:

1. Training, motivational, or informational programs should be carefully designed on the basis of the factors brought to light in the PNA. Specifically, these include audience analysis, objectives, intended use of the program, and so on. Designing a program *without* considering these things is like building a bridge without knowing how long it will be, how many cars it must hold at one time, or how deep the river is that it will span.
2. Loading up a program with wall-to-wall facts, figures, and procedures is a waste of money. Audiences simply cannot absorb large amounts of detailed information presented in a short period of time. Once they begin to realize that they will never remember most of what they're seeing, they will simply turn off and let it all go in one ear and out the other.
3. Audiences tend to remember the things that interest them and that they can relate to. For example, a front-line craft worker will be more likely to retain a message about company cost reductions if it is presented in terms of her job rather than as a series of bar graphs or a list of statistics.
4. The client should look upon the producer as an expert in a very complex field. The producer should see the client as a valued customer and perhaps as a content expert with a very important message that needs to be communicated. Both should understand that what is best for the *audience* is what should determine their course of action. This attitude fosters a sense of mutual respect that allows both parties to do what they do best and usually results in the production of an effective program.

CLIENT COMMANDMENTS

Although the producer should do everything possible to guide the client into a certain frame of mind, there is also a fine line that, when crossed, can have the opposite effect. In essence, it can convince the client that the television department is insensitive to his or her needs and, therefore, is a waste of the company's time and money. Obviously, the fewer of these types of clients the television department has floating around the company, the better off it is—especially if they are influential.

In most cases, it is the producer who must walk the fine line between trying to give the client the best possible program while at the same time being careful not to turn off the client by squelching his or her ideas, wants, or needs. When interacting in these situations and, for that matter, in any other client interactions, the producer should keep the following commandments in mind.

1. Never let a client feel that you do not want very much to satisfy his or her communication need, even if you cannot do so. Knowing that you care will give the client confidence and make her your friend even if she is eventually referred to a different department for a solution to her problem.
2. Treat a client the same way you would treat a paying customer who has approached you as an independent businessperson to provide a product or service. This independent business approach tends to emphasize the importance of clients because independent businesses, like television departments, are usually survival-oriented. Carelessness in this area can have much the same result as crossing the fine line mentioned earlier. The client who feels mistreated becomes an enemy, broadcasting the bad news about your television department throughout the company.
3. Clients almost always respect a producer who has definite opinions about program production more than one who simply agrees with whatever the client wants. And client respect almost always leads to trust as well.
4. Although the ability to interact well with clients is an important trait for a producer, the most important element is simply a job well done.

Survival in corporate television, then, is not very different from survival in any other type of corporate work. You must be able to play the game when necessary, guide your clients properly but stay keenly aware of when you're pushing them too far, remain unquestionably loyal to your audiences, and—above all—do high-quality work for a reasonable price every time you produce a project.

appendix

The Production Process Checklist

As we have seen, the production process is quite involved. Since the information we have covered is spread over many pages in this book, the following checklist should provide a source of quick reference for the student or new producer. It covers, in outline form, each of the activities we have explored in the general order in which they appear in the production process.

As I mentioned before, this is not the only order in which corporate television programs can be developed. It is, however, a good, solid order, including all the proper steps—planning, approvals, and so on. Read it carefully, commit it to memory, and try to follow it whenever possible.

PART I: WRITING AND DESIGN

I. Client contact. Meeting is arranged between producer and client.

II. Initial client meeting takes place.

- **A.** Entire production process is discussed with client. This:
  - **1.** Informs client.
  - **2.** Helps client feel secure.
  - **3.** Enlists client's assistance in all phases of production.
- **B.** Information needed to write the program needs analysis (PNA) is discussed:
  - **1.** Problem to be solved
  - **2.** Background information on the problem
  - **3.** Thorough audience analysis
  - **4.** Specific, quantifiable objectives
  - **5.** How the program will be used
  - **6.** Summary recommendation

III. PNA is written by producer.

IV. Decision is made whether or not to proceed with program development. This should be based on the following:
- **A.** Is the problem best solved with visual help?
- **B.** Are the objectives achievable on videotape?
- **C.** Is the time frame achievable?
- **D.** Does the subject have longevity?
- **E.** Does the importance of the message justify the expense?

V. PNA is sent to client for approval; client approves PNA.

VI. The writer is hired.
- **A.** Question résumés.
- **B.** Seek out professionals.
- **C.** Consider other producers as sources.
- **D.** Share PNA with writer and discuss program design.
- **E.** Writer meets with client to:
  - **1.** Make introductions.
  - **2.** Gather information to write the content outline and/or treatment.

VII. Writer produces content outline (if appropriate).
- **A.** Purpose is to organize content.
- **B.** Outline should prove that the writer:
  - **1.** Understands research material.
  - **2.** Is including and excluding the proper facts.
  - **3.** Has found a logical, easily understood order for the material.

VIII. Content outline is sent to client; content outline is approved.

IX. Writer produces treatment.
- **A.** Can be verbal at first.
- **B.** Describes program visually.
- **C.** Incorporates content points that support objectives in PNA.
- **D.** Gives producer idea of time, money, and so on required to produce the program.

X. Treatment is sent to client; client approves treatment.

XI. Writer produces the first draft of the script.
- **A.** Various formats:
  - **1.** Host on camera
  - **2.** Role-play
  - **3.** Host and role-play
  - **4.** Voice-over narrator

B. Various styles:

1. Film style:

a. Uses full page for narration/dialogue and camera directions.

b. Used for single-camera shooting.

2. Television style:

a. Uses two columns—sound and picture.

b. Used for multicamera shooting.

C. Basic script elements:

1. Uses sound and picture wisely

2. Structured for natural flow

3. Simple, conversational

4. Producible

D. Second and third drafts are written and sent to client.

1. Receive input from client.

2. Avoid approval by committee if possible.

XII. Final draft is sent to client. Script is approved.

PART II: PREPRODUCTION

I. Involves organization, scheduling, and confirmation of all elements of production.

II. Usually involves four people:

A. Producer

B. Director

C. Assistant director

D. Client

III. Involves the following tasks:

A. Budgeting:

1. Done by producer.

2. Based on breakdown of time, equipment, and people.

3. Usually most effective at time of shooting script.

4. Budget range established at PNA stage.

B. Script review and breakdown:

1. Done by director or assistant director.

2. Breaks program into lists of elements that need to be arranged—talent, locations, equipment, props, etc.

3. Director may use it as a scheduling tool.

4. Assistant director uses it to begin line-up work.

C. Obtaining props and wardrobe:
   1. May be as simple as a few work tools.
   2. May be complex—wardrobe rental, house, actor's fittings, etc.

D. Scouting and confirming locations:
   1. Director scouts early in preproduction.
   2. Must visualize how scenes will play out.
   3. Must confirm that location will work logistically.
   4. Possible disruption to employee locations:
      a. Should be considered by assistant director. AD obtains permits from city or county offices and private parties.

F. Auditioning and selecting talent:
   1. Arrangements are made by assistant director from pictures, books, other programs, and agencies.
   2. Director makes final selections:
      a. Input from producer is common.
      b. Client may also want input.
      c. Avoid casting by committee.
      d. Employee selections should involve interview with director.

G. Developing the shooting schedule:
   1. Director's job after three basic determinations:
      a. Best shooting order
      b. Time needed for each shot
      c. Most economical use of talent, locations, and crew
   2. Should be reviewed and approved by the producer.

H. Hiring crew:
   1. Usually done by assistant director.
   2. Director is usually most interested in director of photography.
   3. Typical crew for location shoot:
      a. Director
      b. Assistant director
      c. Director of photography
      d. Engineer-in-charge
      e. Gaffer
      f. Grip
      g. Teleprompter operator
   4. Typical crew for studio shoot:
      a. Director

- **b.** Assistant director
- **c.** Technical director
- **d.** Director of photography and camera 1
- **e.** Camera 2
- **f.** Camera 3
- **g.** Floor manager

I. Building and lighting sets:

- **1.** Can be simple, requiring only several hours to build.
  - **a.** Director of photography is in charge.
  - **b.** One or two grips construct sets.
- **2.** Can be complex, requiring several days:
  - **a.** Art director designs sets.
  - **b.** Set builders construct sets.
- **3.** Sets are completed just prior to production.

J. Renting and reserving production equipment:

- **1.** Assistant director makes arrangements.
  - **a.** Rental facility delivers.
  - **b.** Production assistant or crew member picks up.
  - **c.** If equipment is owned, assistant director reserves it for shoot dates.
- **2.** Typical location shoot equipment checklist includes:
  - **a.** Camera and power supply
  - **b.** Head, tripod, and spreaders
  - **c.** Recorder
  - **d.** Video monitor
  - **e.** Waveform monitor
  - **f.** Various microphones
  - **g.** Audio mixer
  - **h.** Two light kits with gels and diffusion
  - **i.** Two reflectors
  - **j.** Videotape
  - **k.** Video cables
  - **l.** Audio cables
  - **m.** A/C cables
  - **n.** Generator/assorted batteries
  - **o.** C stands
  - **p.** Flags and scrims
  - **q.** Gaffer's tape
  - **r.** Dulling spray

s. Makeup kit
t. Slate
u. Felt tip markers

K. Designing and creating artwork:
1. Assistant director handles it.
2. Usually consists of:
a. Graphs/lists
b. Special slides
c. Character-generated titles
3. Lead time can be a factor.

L. Blocking the script:
1. Director's job.
2. Overhead sketches or storyboards:
a. Primarily done in preproduction.
b. Often fine tuned in production.
c. Dramatic scenes should be finalized in production or rehearsal.

M. Rehearsals:
1. Rare in corporate television
2. Good idea if program is dramatic and involves role-play.
3. Technical rehearsal, usually day before studio shoot, to run through all production elements—angles, lighting, audio, and so on.

N. Preproduction meetings:
1. The more complex the program, the more meetings required.
2. One meeting usually adequate in corporate television:
a. Held one day before the shoot.
b. Scripts and shooting schedules handed out to crew.
c. Director reviews all shots.

O. Preparing the equipment:
1. Done by maintenance engineers on a regular basis.
2. Done just prior to the shoot.

PART III: PRODUCTION

I. Basic formats:
A. One-inch reel-to-reel
B. Three-quarter-inch U-matic
C. Half-inch camcorder

II. Involves two types of shooting:

A. Single-camera—requires camera to be moved for each angle recorded.

1. Action is repeated.

2. Often, shot list includes wide shot, medium shot(s), and close-ups.

3. Editor later assembles piece by piece.

B. Multicamera—means several cameras, which director takes as performances play out.

1. Action need not be repeated for different angles.

2. Editor later puts together chunks of pre-edited footage.

III. Typical location (single-camera) shoot

A. Crew arrives at location.
B. Grips and gaffers unload equipment.
C. Director and director of photography discuss first shot.
D. Assistant director works with actors and clients.
E. Engineer-in-charge sets up monitors, recorder, and sound.
F. Grips/gaffers set lights per instructions from director of photography.
G. Teleprompter operator sets up equipment on camera.
H. Rehearsal
I. Tweeking lights, moves, action, and so on
J. First take
K. Additional takes
L. Director buys a take.
M. Assistant director notes buy.
N. Camera is moved for additional coverage.
O. Location is struck, crew moves to next location and sets up.
P. Process is repeated for all remaining scenes.

IV. Typical studio (multicamera) shoot:

A. Crew call approximately one hour before talent.
B. Cameras set up, white-balanced, registered, etc.
C. Actors called to set.
D. Rehearsal
E. Tweeking and adjustments
F. First take
G. Adjustments
H. Director buys a take.
I. Assistant director circles buy.
J. Process is repeated for all remaining scenes.

V. Audio production—studio or location:

A. In studio:

1. Takes place in production audio.

2. Recorded on quarter-inch audio tape.

3. Transferred later to videotape.

B. On location:
  1. If possible, in same general area as video recording.
  2. Recorded on video tape, although pictures are not used.

PART IV: POSTPRODUCTION

I. The postproduction process:
  A. Duplication:
    1. Duplicate originals
    2. Work prints; window dubs
  B. Development of master script package:
    1. Master script and all other notes
    2. Window dubs of original footage
  C. SMPTE Time Code:
    1. Code is standardized by the Society of Motion Picture and Television Engineers.
    2. Electronic signals are recorded on cue or address track.
    3. Code appears visually in a window dub as segments of time.
    4. Editor uses code to make electronic transfers of video and audio segments.

II. The off-line process:
  A. Editor first meets with producer for input, then reads script to get a feel for pacing and tone of program.
  B. Editor views window dubs of footage to get familiar with footage and notes/shot report.
  C. Editor loads a stripped tape into the record machine.
  D. Editor loads window dub of reel with first shot into playback machine.

III. Uses master script and videotape shot report to move through footage and finds the circled take (buy) of first shot.
  A. Selects in point and out point on playback footage (material to be recorded).
  B. Selects in point and out point on record tape (place where material will be recorded).
  C. Presses preview button on controller to see the edit without actually performing it.
  D. Makes adjustments to edit.
  E. Presses perform button. Controller now:
    1. Prerolls both machines to five seconds prior to edit points.
    2. Rolls both machines at identical speeds (in sync).
    3. Records the footage between the in and out points on the playback machine onto the in and out times of the record machine.
    4. Ends recording at the out times specified.

IV. If no list management function, editor now writes time code numbers corresponding to edits, starting a list.

V. If list management function is part of system, time code numbers are automatically stored on disk.

VI. Process is repeated for every edit (event) in the program.

VII. When off-line or rough cut is complete, editor gives back to producer:
- **A.** Master script package
- **B.** Window dubs of original footage
- **C.** Completed rough cut
- **D.** List or disk

VIII. The client screening:
- **A.** Explain that program is a rough shot-by-shot assembly, not a polished version.
- **B.** Explain that this is the last chance to make changes.
- **C.** If possible, include editor and director, who are familiar with the footage.
- **D.** Carefully note any changes.

IX. Decision to move ahead to on-line or make changes in additional off-line work depends on:
- **A.** Is on-line being rented or is it owned (rental costs are often prohibitive)?
- **B.** Are changes numerous or minor?

X. Client approves final off-line.

XI. The on-line edit:
- **A.** Original footage and first-generation duplicate footage now used; time code is identical to window dub copies.
- **B.** Usually a much more sophisticated editing system.
  - **1.** Controls numerous videotape players/recorders.
  - **2.** Interfaces with switchers, audio board, and special-effects generators.
  - **3.** Capable of auto assembly.
    - **a.** Editor enters list from off-line.
    - **b.** Gives auto-assemble command.
  - **4.** Normally auto-assembled in sections.
  - **5.** Special effects added.
  - **6.** DVE (digital video effects) added.
  - **7.** Character-generated titles added.
  - **8.** Minor sound mixing may also be done.

XII. On-line edit is complete.

XIII. Audio sweetening:
- **A.** Purpose is to enhance the original audio track.
- **B.** Done in a sweetening room:
  - **1.** Audio channels from on-line (edit master) laid down onto multichannel audio tape.

2. Sound modified and additional sound effects, music, etc., added on separate channels.
3. When satisfactory, multiple channels are mixed down on one or two channels.
4. Final mixed channel is laid back onto the edit master in sync with pictures.

XIV. Program is now complete; final copy is duplicate master, which can be used to make distribution copies for the client's audience.

XV. Evaluation:

A. Necessary to:

1. Gauge audience acceptance of program.
2. See effectiveness of program in meeting objectives.
3. Provide proof of television department's value to company.

B. A good evaluation should:

1. Tell the producer if the program accomplished the objectives established in the PNA.
2. Tell the producer if those objectives were the right ones to help solve the problem on which the program was focused.
3. Tell the producer how acceptable the program was aesthetically to the audience.
4. Give the producer input on ways to improve areas that fell short of audience expectations.
5. Be simple enough not to intimidate the audience or person administering it.

C. Administering the evaluation:

1. Can be handled by mail, in person, or by a presenter.
2. Good for producer to be present at some evaluation sessions to gain verbal input also.

D. After evaluation is complete:

1. Tally results.
2. Look for patterns of likes or dislikes.
3. Make changes in future programs as appropriate.
4. Document results and save for presentation to executives or budget committees.

E. Indices evaluation:

1. Evaluates actual dollar savings to company on the basis of indices kept by client or producer.
2. More credible to executives and budget committees than audience acceptance evaluations.

Glossary

A/C (cable) A/C stands for alternating current and refers to the standard electricity required to power the lights, recorder, camera, and so on. When used in production, A/C also refers to the cables into which the lights and other equipment are plugged.

ambience The background or environmental noise of a particular location. For instance, the ambient noise on a street corner in a big city would be car and bus engine noises and horns honking. Ambient noise for a summer evening exterior scene shot in the country would probably include crickets, frogs, and perhaps dogs barking in the distance.

American Federation of Television and Radio Artists (AFTRA) One of two primary unions to which many actors belong. (*See also* Screen Actors Guild, SAG.) AFTRA members work in television and radio, not the feature film industry.

assistant director (AD) Also called *associate director.* The initials are much more commonly used than the full name. This person is the director's closest assistant and helps him accomplish any and all details, which gives the director the freedom to work on his primary responsibilities.

audio The sound portion of a shot, scene, or production. On location, also used to mean audio cables.

audio sweetening The process by which the final audio enhancement and mix-down is done on a program. The sweetened audio track is laid back onto the edit master of the program as a final step in the postproduction process. Also known as *post audio.*

audition An actor's tryout for a part. Usually set up by the assistant director and attended by the director and producer and, in some cases, by the client as well.

backlight A light directed at an actor from behind, which highlights his or her head and shoulders and serves to make the actor stand out from the background. Also called *rim light.*

barn doors The flat black metal flaps or cutters that are attached to the front of many production lights. Barn doors allow the light to be cut off from certain areas of a set or scene.

bars and tone A video and audio reference used to assure that chroma (color), luminance (light intensity), and audio (sound) are kept consistent between pieces of production equipment and recorded segments of tape. Appears as a series of different colored vertical bars spanning the color range. Tone is a high-pitched, 1,000-cycle tone.

boom An extendable pole on which a microphone is attached. The boomed microphone is held in close relation to actors during a shot. It wires back to the audio mixer and, ultimately, the videotape recorder. Also called a *fish pole.*

breakdown (script) An itemization of each of the elements of a script, which enables an assistant director or production assistant to begin lining up each of those elements for the production itself.

buy A director's term, which tells the crew and cast that the shot just recorded was satisfactory.

C stand A versatile, three-legged metal stand used in production for a variety of purposes, but primarily to hold up flags, silks, and nets.

chroma An engineering term meaning *color.*

chroma key A process by which a picture element is combined with another picture. Chroma keying is usually accomplished by shooting one element, such as a person, in front of a blue background in a studio. A video switcher can then replace the color blue behind the person with a totally different picture source, such as a shot of a downtown city street. Also called a *key.*

circled takes The director's buys, which the assistant director or script supervisor circles in the master script notes. These later let the editor know which are the shots preferred by the director.

client The person who requests the making of a videotape program and who is often the primary approving person for all aspects of the production.

close-up Refers to the focal length and framing of a camera shot. A close-up of an individual would include only his or her face, filling the screen. A close-up of an object would, likewise, fill the screen.

code A term referring to SMPTE Time Code, the digital signals that allow very precise editorial control. (*See also* SMPTE.)

content expert A person, usually designated by the client, who is an expert on the subject covered in the script. The content expert is normally placed at the disposal of the writer. (*See also* SME.)

controller The piece of equipment used in an edit bay to interface the two videotape machines. Allows the editor control over the record, playback, and shuttle functions. In some cases, also provides the automatic list management function.

control track A system of videotape editing in which continual pulses are recorded on the tape by the camera and later used to isolate specific segments for transfer. Control track offers less flexibility than SMPTE Time Code and is thus less desirable.

coverage The number and types of shots a director plans in order to cover a shot or scene adequately. Typical coverage would include a wide or master shot of all the action in the location, medium shots of the individual actors, and close-ups of key people and actions.

crane A platform on a large boom on which the camera, the camera person, and an assistant (or the director) can be seated. Used to get high angle shots during which the camera is raised up or lowered down.

cue A signal to an actor to begin his or her portion of the scene. Cues are also given to crew members to initiate certain actions—for example, to the camera person, to begin a zoom. Also refers to a video or audio source *cued up* to a certain spot for use.

cut Has several meanings. When used by the director in production, *cut* means stop recording: Either the scene has ended or something has gone wrong. Also used to describe an edit in a program—an instantaneous change from one shot to another. Can also refer to a *cut* of the entire program—for instance, the *rough cut.*

digital video effects (DVE) Created by pieces of electronic equipment that are used to modify the video image. These include pictures that shrink away off the screen, are posterized to look like paintings, break up on the screen, and many other visual effects.

director The person in charge of translating the script into effective completed videotape shots and scenes, which an editor can then assemble into a complete program. The director bears the responsibility for every crew and cast member, as well as every aspect of the program while it is in production.

director of photography (DP) The person in charge of seeing that all pictures recorded for the program are properly lit, exposed, and photographed. The director of photography assists the director in recording action in such a way as to create the type of mood and pace required. In videotape production, also called the *lighting director* or *director of videography.*

dissolve A basic editing effect in which one picture begins to fade out while a second begins to fade up. The effect is a soft transistion of momentarily overlapped images. Normally used to create a sense of passage of time between one scene and another.

dolly As a verb, refers to the movement of the camera to follow some piece of action, commonly referred to as a *dolly shot.* As a noun, refers to the piece of equipment on which the camera is mounted in order to carry out that movement—a base or structure on wheels with handles at one end for pushing, turning, or pulling. A dolly often travels on metal rails called *track* for smooth, level movements.

dub Jargon term meaning a duplicate of a program, shot, scene, etc.

dulling spray An aerosol spray that is applied to shiny items seen in a video picture. Dulling spray cuts down on glare and thus reduces hot spots.

dupe master In videotape, equivalent to an internegative in film—a high-quality source (copy of the program) from which distribution copies can be made.

edit Can refer to the modification of a piece of writing, such as a script or treatment. In videotape, refers to the points at which different shots are placed (*edited*) together to create a completed program. Also refers to the *type* of edit—a dissolve, a cut, a wipe, etc.

edit master The final edited version of the program, which results from the on-line editing session. The edit master is created from the original footage and first-generation duplicates as much as possible for maximum sound and picture quality.

editor The person responsible for assembling an off-line or on-line copy of the program from the original footage or duplicates.

engineer-in-charge (EIC) The head engineer on a studio or location shoot. Typically, the EIC is responsible for the operation of all electronic equipment. Usually works closely with the director of photography to establish proper video levels. Also works with the sound engineer to assure proper audio levels, or may double as the sound engineer or technical director.

executive producer Typically, the person providing or arranging for the financing needed to produce the program. In corporate television, usually the department head or manager for whom the producer works. Oversees the quality and cost-effectiveness of the producer's work and is directly accountable for all or part of the department's budget.

false start Term used in production when something goes wrong in a scene just after it has begun. When the director or assistant director calls out, "False start," this indicates that a new slate and scene number are not needed. Some signal is usually made for the editor, however, such as the cameraman waving his hand in front of the lens for a moment. This means a new start will follow.

flag A piece of black cloth stretched over a metal frame, used to cut off light from an object or actor. Usually placed on a C stand or hung from a lighting grid in the studio in order to accomplish this.

fill light Refers to the use of, or need for, a fill light to dim or fill in the shadows created by the key light.

focus puller A person who assists the camera person in making a follow focus move. The focus puller "pulls," or adjusts, focus while the camera person is busy panning, tilting, zooming or a combination of the three.

follow focus The need to change focus as the camera and/or actors are moving during a shot.

gaffer An electrician, lighting specialist, and key assistant to the director of photography or lighting director.

gaffer's tape Very strong tape, usually gray in color, which is used in production for a variety of purposes.

gels Thin pieces of colored film placed over light sources. Gels change the color of the light being emitted and in some cases correct it to a specific color temperature.

grip The person whose primary job is to load and unload or move equipment and cables. A grip usually works under the gaffer, helping to build and strike lighting and camera setups.

handheld Footage shot with the camera placed on the camera person's shoulder or in his or her hands instead of on a tripod or dolly. News crews often shoot handheld in the interest of time. Production crews shoot handheld either for effect or to facilitate camera movement or shooting in tight quarters.

host or hostess An on-camera person who talks directly to the audience (i.e., the camera), usually leading the audience through the material to be conveyed.

insert Usually a close-up or extreme close-up of something *inserted* into the main action. For example, the host, seen in a wide shot, points at a tool. As he does so, we cut to an insert of the tool to focus attention on it. When the host finishes pointing, we return to the wide shot.

key (light) The main source of light used to illuminate a scene. The key light is usually placed close to the camera and shines down on the actor(s) at approximately a 45 degree angle. *Key* is also an abbreviation of *chroma key.*

lavalier Usually a small, condenser-type microphone often placed on a tie clip or easily camouflaged under a collar or lapel.

location Any place where a film or videotape shoot is taking place *other* than in the studio.

luminance An engineering term meaning light intensity.

manual list The numeric edit decision list written down by the off-line editor. The manual list shows the time code numbers corresponding to each edit he or she has performed.

"Mark it" A request for the person holding the slate to speak the scene and take number into the microphone. This serves to *mark* the beginning of the scene for the editor.

master script The shooting script on which the assistant director or script supervisor notes which scenes have been covered and how. Her notes include brief shot descriptions and the director's reasons for liking, disliking, or "buying" those shots.

master shot A recording of the shot, usually with all actors and actions visible and the setting established. Typically, a master shot begins a scene and is recorded in a wide-angle focal length.

medium shot A shot recorded in a focal length somewhere between a wide-angle and a close-up. Typically features one or two people, showing them from about the waist up.

mixer *Sound mixer* refers to the person responsible for the point of input of all sound sources. He or she *mixes* the sound to the proper levels as it is being recorded or *sweetened.* A *mixer* is also the piece of equipment used by the sound mixer—a point of termination for all sound sources, with controls to change the volume of individual sources as needed for a smooth mix.

monitor A high-quality version of a television without the tuner for channel selection. A monitor has controls allowing for precise picture tuning and clarity from sources such as cameras. It also has video and audio inputs compatible with production cabling and hardware.

narrator A voice that narrates the program we are watching. Unlike a host or hostess, a narrator never appears on camera.

off-line The first edited version of a program. The off-line is a work print done with minimal editing equipment, used as a shot-by-shot assembly of the program for review and preliminary approval purposes. Also called a *rough cut.*

on-line The final edited version of a program. Differs from an off-line in that it is usually done with an expensive computerized editing system capable of autoassembly and various special effects. The on-line also uses the original footage and, whenever possible, first-generation duplicates. Also called a *fine cut.*

original footage The footage brought back from the field. Original footage will have the highest quality picture and sound. Quality begins to deteriorate with each generation of duplicates.

over-the-shoulder (OS) A shot in which the camera is looking over one person's shoulder while actually featuring or favoring another person.

pan A horizontal camera movement following an action sequence or revealing something to the left or right of the original framing. *Pan* is short for *panorama.*

producer The person who guides and is accountable for all aspects of the production process, from the time the client requests a program until the program is duplicated and sent to the field for use.

production assistant (PA) A person usually brought in by the assistant director or producer to handle duties such as phone calls, paperwork, or deliveries in preproduction or production.

program needs analysis (PNA) A short analysis of certain basic information such as the client's needs, audience knowledge, and so on. The PNA is usually written by the producer, and is used to determine whether a videotape is the proper tool for the job and, if so, what basic design the program should take.

reflector A shiny reflective surface (usually hard and boardlike) used to bounce sunlight onto a subject as a lighting source.

registration An engineering term meaning that the red, blue, and green images produced by a video camera are in proper alignment—perfectly superimposed over each other.

role-play A production or part of a production in which a plot is introduced and actors play character roles. The role of host is considered a spokesperson rather than a character.

"Roll-tape" The cue given by the director or assistant director to tell the engineer-in-charge or technical director to begin recording a scene.

Screen Actor's Guild (SAG) One of two primary unions to which many actors belong. (*See also* AFTRA.) SAG is the primary union for performers in the feature film industry.

scrim A wire mesh insert, usually circular, that is placed into a lighting instrument to diffuse or lower its level of intensity. Scrims are numbered by density to produce a predictable effect on the camera's f/stop settings.

script The written foundation or skeleton of the program, containing all dialogue and camera directions. Its importance to the production of a good program cannot be overstated. Although the director is usually free to take some liberties with the script, it should be followed as closely as possible.

setup The setup for a shot consists of camera placement, all lights, flags, and so on. Also used to mean camera setups only. The term *setup* also refers to the setup of certain pieces of production equipment, such as waveform monitors and vectorscopes, which ensure proper and consistent color and light intensity levels between different recorded or broadcasted segments.

shooting schedule The itemized shot-by-shot schedule outlining the entire production. Includes all scene numbers, locations, time required to get each shot, and any important notes regarding a particular shot. Usually written by the director and approved by the producer.

shotgun A long, slender microphone usually used on the end of a boom. Its pick-up pattern allows it to record sounds in a conelike area directly in front of it, but it picks up very little sound behind it.

shot list The list of all shots required to cover the material in the script. Written by the director.

silk A large, thin piece of white fabric stretched onto a metal frame and placed above the actors in a shot to remove harsh shadows resulting from direct sunlight.

single A shot of a single actor, often used synonymously with *medium shot* or *medium close-up.*

slate The small board held in front of the camera at the beginning of each scene and used to communicate scene information to the editor. In videotape, the wooden clapper is not used, since time code is the sync source. In film, however, the clapper is used to create an audio sync mark between the sound track and the visuals.

Society of Motion Picture and Television Engineers (SMPTE) The association that standardized SPMTE Time Code. (*See also* code.)

"speed" The cue given by the technical director or engineer-in-charge to let the director know the videotape and/or sound recorder is up to speed and in record mode.

strike Jargon term meaning to dismantle a setup or an entire set once production is complete at that location.

subject matter expert (SME) Another term for *content expert.* Often acts as a technical advisor on the script or during production.

switcher An electronic interface device capable of switching video sources instantaneously as cuts or with the use of various special effects such as wipes, dissolves, and fades. Used primarily in studio production and on-line editing.

talent Actors, either professional or company employees.

tape op. Short for *videotape operator.* A tape op. is normally used in editing and studio productions that require a remote videotape machine to be rolled at a specific time. This source is then taken by the technical director at the switcher and is thus routed to the production recorder or outgoing broadcast circuit.

technical director (TD) The person responsible for rolling and stopping tape in the control room as well as switching a live broadcast or multicamera shoot.

teleprompter A device that is mounted on the front of a camera and projects the actor's lines on a translucent pane of glass positioned in front of the camera's lens. Allows the actor to look directly at the camera and read the lines.

tilt A vertical movement of the camera's view in a seesaw motion. In a *tilt up,* the rear of the camera is lowered as the lens moves upward. Such tilt would be used, for example, to start on a shot of a building's entrance and slowly tilt up to reveal a man standing on a ledge on the fifth floor. A *tilt down* is the opposite.

time base corrector (TBC) A device used to control and enhance the stability of a video picture.

time code generator An electronic device that generates time code for recording on video and sometimes audio tape.

treatment The first visualization of the script developed by the writer. Used to give the producer and client an explanation of how the writer plans to develop the script.

truck A horizontal movement of the camera, normally when on a dolly.

two shot Usually a medium shot in which two actors are seen.

vectorscope An electronic scope used to align precisely and assure consistency and accuracy of the colors in a video picture.

video The visual part of a program only. Also used as a very general term for a videotape program or the entire video field.

voice over (V.O.) A voice recorded, usually in a sound booth, and heard over a series of video pictures or an entire program. Voice over is considered narration, spoken, of course, by a narrator.

waveform monitor An electronic scope used to measure and adjust various elements of a video signal. In production, the waveform is used primarily to assure consistency of the luminance (light intensity) of a video picture.

wide shot A wide angle camera shot in which all actors and action are visible and recorded. Usually used as an establishing shot and/or master shot.

zoom A change of focal lengths from wide to close or vice versa on a special lens designed for such use.

zoom lens A lens that allows changes of focal lengths either by sliding or rotating an external metal ring similar to a focus ring or with the use of a small electric motor that attaches to the lens.

Bibliography

WRITING AND DESIGN

Matrazzo, Donna. *The Corporate Scriptwriting Book.* Portland, OR: Communicom, 1985

Swain, Dwight V. *Scripting for Video and Audio-Visual Media.* Stoneham, MA: Focal Press, 1981.

Swain, Dwight V., with Joye R. Swain. *Film Scriptwriting: A Practical Manual,* 2nd ed. Stoneham, MA: Focal Press, 1988.

Van Nostran, William. *The Nonbroadcast Television Writer's Handbook.* White Plains, NY: Knowledge Industry Publications, 1983.

PREPRODUCTION

Van Deusen, Richard E. *Practical AV/Video Budgeting.* White Plains, NY: Knowledge Industry Publications, 1984.

Wiese, Michael. *Film and Video Budgets.* Stoneham, MA: Focal Press (Westport, CT: Michael Wiese Film Production), 1984.

PRODUCTION

Alkin, Glen. *Sound Recording and Reproduction.* Stoneham, MA: Focal Press, 1981.

Alten, Stanley. *Audio in Media,* 2nd ed. Belmont, CA: Wadsworth, 1986.

Armer, Alan A. *Directing Television and Film.* Belmont, CA: Wadsworth, 1986.

Cartwright, Steve R. *Training with Video: Designing and Producing Video Training Programs.* White Plains, NY: Knowledge Industry Publications, 1986.

Eargle, John. *The Microphone Handbook.* Indianapolis, IN: Howard W. Sams & Company, 1982.

Fuller, Barry, et al. *Single-Camera Video Production.* Englewood Cliffs, NJ: Prentice-Hall, 1982.

Gayeski, Diane M. *Corporate and Instructional Video: Design and Production.* Englewood Cliffs, NJ: Prentice-Hall, 1983.

Hickman, Walter A. *Time Code Handbook.* MA: Datametrics–Dresser Industries, Inc.

Huber, David Miles. *Audio Production Techniques for Video.* Indianapolis, IN: Howard W. Sams & Company, 1987.

Jacobs, Bob. *How to Be an Independent Video Producer.* White Plains, NY: Knowledge Industry Publications, 1986.

Le Tourneau, Tom. *Lighting Techniques for Video Production.* White Plains, NY: Knowledge Industry Publications, 1987.

Medoff, Norman J., and Tom Tanquary. *Portable Video: ENG and EFP.* White Plains, NY: Knowledge Industry Publications, 1986.

Millerson, Gerald. *Technique of Lighting for Television and Motion Pictures,* 2nd ed. Stoneham, MA: Focal Press, 1982.

——. *Technique of Television Production,* 11th ed. Stoneham, MA: Focal Press, 1985.

——. *Video Production Handbook.* Stoneham, MA: Focal Press, 1987.

Nisbett, Alec. *The Use of Microphones,* 2nd ed. Stoneham, MA: Focal Press, 1983.

Oringel, Robert S. *Audio Control Handbook,* 6th ed. Stoneham, MA: Focal Press, 1989.

Wiegand, Ingrid. *Professional Video Production.* White Plains, NY: Knowledge Industry Publications, 1985.

Zettl, Herbert. *Television Production Handbook,* 4th ed. Belmont, CA: Wadsworth, 1984.

POSTPRODUCTION

Anderson, Gary H. *Video Editing and Post-production: A Professional Guide,* 2nd ed. White Plains, NY: Knowledge Industry Publications, 1988.

Browne, Steven E. *Videotape Editing: A Postproduction Primer.* Stoneham, MA: Focal Press, 1989.

Hubatka, Milton C., et al. *Audio Sweetening for Film and TV.* Blue Ridge Summit, PA: TAB Books, 1985.

Schneider, Arthur. *Electronic Post-production and Videotape Editing.* Stoneham, MA: Focal Press, 1989.

PERIODICALS

AV/Video. Torrance, CA: Montage Publishing.

Corporate Television: The Official Magazine of the International Television Association. New York: Media Horizons, Inc.

Corporate Video Decisions. New York: NBB Aquisitions Company.

EITV (Educational-Industrial Television): The Techniques Magazine for Professional Video. New York: Broadband Publications.

In Motion: Film and Video Production Magazine. Annapolis, MD: Steven LeHuray.

Videography. New York: United Business Publications.

Video Manager. Torrance, CA: Montage Publishing.

Video Systems. Overland Park, KS: Intertec Publishing Corporation.

Index